The
Willamette River
Field Guide

Travis Williams

Timber Press
Portland | London

Page 2: Swift shallow water flows through ever-changing gravel bars near Beacon Landing.

The maps in this guide are not intended for river navigation. The Willamette Water Trail Guide provides more detailed maps (available at www.willamettwatertrail.org). Although a map can provide excellent detail and point you in the right direction, river conditions and routes can differ from one season to the next. Channels may have reduced flows or be blocked by woody debris, forcing a change of approach. Thus, although a map can be of great value, trust your eyes. The river channel and conditions at greenway parks and other natural areas can change over time. It is up to you to be aware of your surroundings and the potential dangers that may exist.

The Haseltine Building
133 S.W. Second Avenue, Suite 450
Portland, Oregon 97204-3527
www.timberpress.com

2 The Quadrant
135 Salusbury Road
London NW6 6RJ
www.timberpress.co.uk

Printed in China

Library of Congress Cataloging-in-Publication Data

Williams, Travis.
    The Willamette River field guide / Travis Williams. — 1st ed.
        p. cm.
    Includes bibliographical references and index.
    ISBN 978-0-88192-866-2
    1. Rivers—Recreational use—Oregon—Guidebooks. 2. Willamette River (Or.)—
Guidebooks. I. Title.
    GV191.42.O7W55 2009
    917.95'30444—dc22
                                                                2008032992

A catalog record for this book is also available from the British Library.

*For my parents, Gary and Terra,
who always encouraged my interests,
for Grandma Jan, who helped foster a love of nature,
for Grandpa Joe and Grandma Esther who got me out
in the boat on many days, for Eleanor and Frances
who are a constant joy, and for Sandra.*

Contents

Color photographs follow page 112.

Maps

Preface

WHEN WE HEAR about Oregon's natural beauty, the Willamette River is seldom mentioned. Even people who live quite close to the river look west to the coast or east to the mountains or high desert to find the places that are emblematic of the state's natural gifts. The Willamette River tends to be thought of only in terms of its water quality. Yet the state's natural feature that has been most abused and neglected provides an abundance of natural beauty that is largely undiscovered.

The Willamette River has been greatly affected by the presence of people for more than a century. We have sought to block its flow, tame its meandering course, and harness its water for drinking and industrial purposes. We have used its huge volume to send our pollution down to the next town or the next river. In some ways, it is a testament to the Willamette's resilience that it can be used at all today.

The Willamette holds many surprises. If you explore the river's changes through time and its story today, you'll find that the river is a fundamental aspect of Oregon's history and its future. Today the Willamette's riverside greenway parks, gravel bars, and islands are home to a myriad of wildlife. Going down to the river's edge to see the current and to witness a great blue heron or a beaver on the water rewards the visitor many times over.

Through my work with Willamette Riverkeeper, I've had an opportunity to take a good long look at the river. What I've found has led me back again and again. So take a few minutes or a few days and get to the river. I hope that you will experience its magic, glimpse the amazing array of wildlife that thrive here, or connect to the river in some other way.

See you on the river.

Acknowledgments

I want to thank the great board, staff, and volunteer members of Willamette Riverkeeper who have taught me a lot over the years and provided me an opportunity to really get to know the Willamette. I have also been fortunate to work with and learn from some fantastic colleagues within the state and federal agencies, nonprofit organizations, and other organizations working to improve the condition of the Willamette. Photographer Michael Wilhelm is much appreciated for his helpful suggestions that enabled me to capture portions of the Willamette photographically. I also want to thank the many people I've had a chance to travel the Willamette with over the years whose insight, skill, and vision have helped shape my view of river exploration and conservation, including Karl Adams, Joe Coffman, Barbara May, Amy Morrison, Kurt Renner, Russ and Beverly Smith, Russ Woodward, and Bill Young, to name a few. Also thanks to the Lazar Foundation and the Willamette River Fun(d) of the Oregon Community Foundation for their early financial support of this project. Thanks also to Lisa Brousseau for her commitment to this project which made it much better, and thanks to Jason Clark for his excellent cartography. My appreciation to Richard Engeman for providing insight into the historical photos. And a huge thanks to editor Eve Goodman at Timber Press for her early and ongoing interest, her immense patience, and her dedication to the final product. Finally, thanks to Elizabeth Grossman who had a common interest in this project and helped to get the whole thing launched.

Introduction to the Willamette

A man is rich in proportion to the number
of things he can afford to leave alone.
—Henry David Thoreau

There is a place in the Willamette Valley where you can go on a spring day and immerse yourself in the natural world. As you guide your canoe toward a band of green at the river's edge, you see wildflowers in a grassy opening, migratory songbirds zipping through the trees, and deer standing at the water's edge surveying you. The gurgle of current can be heard here and there as you travel quietly onward. Cottonwoods tower overhead as a beaver plunges into the water from the bank, soon surfacing to get a look at you. This place of green and blue quiet is the Willamette River.

Everywhere along the river wildlife can be seen and heard. Eagles and ospreys live here. Deer, beavers, bobcats, river otters, mink, and others eat at the river's edge. Oaks and camas loom in meadows next to the river. Similar scenes exist all along the river's extent. You can find it in Eugene, Springfield, Corvallis, Albany, Salem, Keizer, Oregon City, and Portland. Other scenes tell a different story as well. From a rich past of massive floods in spring, low summer flows, and a long connection to the people of the region, it does not take long to understand that the Willamette River is a complex creature. You might even say that the river is a living organism, one that thrives on the basics of life, such as clean cold water and rich riverside lands.

The Willamette is the thirteenth largest river in the United States, based on the annual volume of flow. The average flow at the Willamette's confluence with the Columbia River is 32,000 cubic feet per second (cfs). Normal average flows range from 8000 cfs in August to 70,000 cfs in December, but the river is estimated to have reached 460,000 cfs in the February 1996 flood. The river stretches 187 miles on the mainstem, which begins just southeast of Eugene and flows from south to north. The Willamette River Basin encompasses 11,478 square miles and contains thirteen significant tributaries: Coast Fork Willamette, Middle Fork Willamette, McKenzie, Long Tom, Marys, Calapooia,

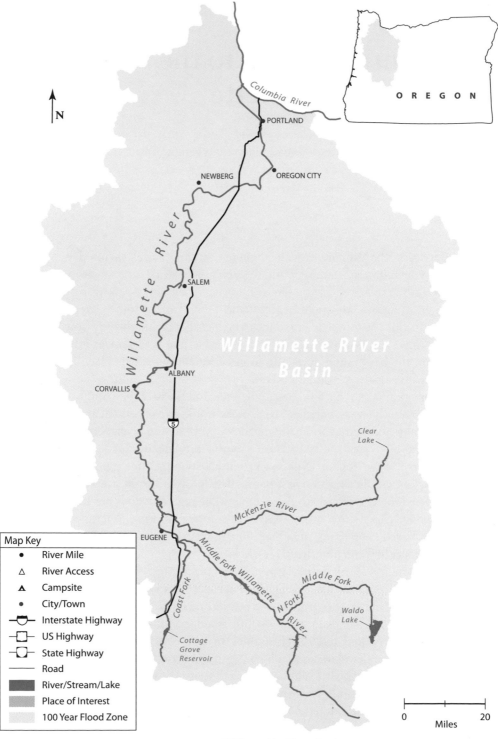

N

OREGON

Columbia River

PORTLAND

NEWBERG   OREGON CITY

Willamette River

SALEM

Willamette River
Basin

ALBANY

CORVALLIS

Clear
Lake

5

McKenzie River

EUGENE

Middle Fork Willamette   Middle Fork

Coast Fork   N Fork

River

Waldo
Lake

Cottage
Grove
Reservoir

Map Key

•   River Mile
△   River Access
▲   Campsite
•   City/Town
⬡—   Interstate Highway
▢—   US Highway
▢—   State Highway
——   Road
▬   River/Stream/Lake
▬   Place of Interest
▬   100 Year Flood Zone

0   20
Miles

Willamette River Basin

Luckiamute, Santiam, Yamhill, Mollala, Pudding, Tualatin, and Clackamas. Approximately 2.4 million people live in the Willamette Valley, and this figure is projected to increase to 4 million by 2050. At present, the valley accounts for 70 percent of Oregon's population.

The Willamette River is fed by a rich tapestry of lands. It is bordered by the Cascades to the east and the Coast Range to the west, with foothills nestled against their flanks. Overhead the clouds loom and the mist drifts through the trees, feeding the mountain streams with water. The streams rush ever downward, dropping ice cold water from the Cascades and Coast Range in a steep descent toward larger flows far below. The water flows underground and across the high grassy slopes, sustaining the flows that make the Willamette and its tributaries live year-round.

Patterns can be seen from the heights overlooking the valley—the circle of crops shaped by irrigation equipment on the Willamette's rich floodplain, rectangles hewn by tractors and fences interlocking in all sizes and configurations. The geometry of these agricultural enterprises expands across the valley, much of it to the river's edge. In the cities and towns, we see the weave of roads, with a vast expanse of houses and buildings. Creeks and wetlands shift against neighborhoods, farms, and roads, and many of these eventually flow toward the Willamette. In all of this patterned landscape shaped by people, natural things maintain a presence—in some places rich and in others just holding on.

A few strongholds of riverside willows, cottonwood, and ash remain. Here and there the old channels maintain their shape, receiving high flows that spread into and across large bottomland areas. More often the dynamic flow is held in check by rock and wood, a primitive but effective armoring of the riverside. Overlain on this foundation of moving water, ancient rock, and mud is the never-ending movement of people. Highways crisscross the flatlands and extend back and forth across the current. Cities push and pull at the riverside, from Eugene and Springfield in the south to Portland in the north. Yet, for many of us, the Willamette River is known only in name.

Although the Willamette is the defining natural feature in the nearly 12,000 square miles of the Willamette Basin, those who travel the valley on a regular basis may be better acquainted with Interstate 5 than with the river. While unfortunate, this underscores the level of anonymity that the Willamette possesses. One might go so far as to say that the river hides in plain sight. Many residents of the Willamette Valley see the river on a regular basis. We drive over it every day and look down at its dark water far below. We may even think a bit about the river as we pass over it, perhaps reflecting on its history and present condition. The image of the Willamette, there in our town or along that quiet country road nearby, is perhaps all that valley residents really know about it.

The Willamette River is a mosaic, with many pieces fitting together in a pattern that is both distinct in certain areas and showing similar patterns across its extent. The river travels over 20 miles through Oregon's largest city, Portland, and its suburbs. In the Portland area the Willamette is wide and deep and is hemmed in by industrial facilities and urban buildings. In this stretch, the river is perceived as dirty and in need of a cleanup due to the city's 150 years of industrial development. Although Portland Harbor is in dire need of a cleanup of contaminated sediment, this is not true of the whole river. In Salem, the river is tranquil and interspersed with gravel bars and islands as it flows past the capitol. Further south in Eugene, the Willamette is a shallow river with occasional rapids that pulse onward under the Interstate 5 bridge. Residents of Eugene understand that, while their stretch of the Willamette is not highly impacted by industrial development, the city's growth has its own consequences and protecting the river's water quality in the upper part of the valley is critical. Each of these views of the river may be accurate for that particular stretch, but they do not reflect the river throughout its entirety. However, the issues in each of these stretches do have some bearing on the health of the whole river system.

The Willamette River's history is complex, but as for many other rivers people have a close connection to its story. In some places people have fought what the river does, building dams and concrete embankments to control its channel and seasonal flooding, while in a few others they have sought a kind of balance with what the Willamette wants to do naturally. Over the course of human connection to the river, it has been used to service numerous needs. The river has provided drinking water and numerous fish and plants, such as lamprey, salmon, and wapato. Its annual floods have enriched the low-lying riverside lands used for farming. Prior to the establishment of the railroad and major roads, the Willamette served as a means of travel and commerce between communities. The river has been dammed for hydropower and used as a convenient place to dump waste. In the years since French Canadian and American settlers arrived in the valley, the Willamette has most often been viewed, first and foremost, as a resource to be utilized. Over time we have grown accustomed to taking from the river.

The consequences of this have been hard on water quality and the river's native fish. In our acquisitive ways, we've taken away most of the river's dynamic nature, cutting off its side channels and relegating the river to usually just one main channel, far from its long-held vitality. We've simultaneously utilized areas that were historically enriched by the river's floods, commonly known as floodplains, and set them to other uses. When people build homes and other structures in these low-lying areas, they sometimes have been surprised when the areas still occasionally flood.

While the river has significant environmental issues that affect its health, today Oregonians are learning how to *give back* to our river. Giving back to the Willamette is something that all of us can do, whether through efforts of habitat restoration, the recovery of native fish species, or reducing pollution in urban areas and places where the river's historic habitat and tributaries have been significantly degraded. There is an opportunity for each of us to make a difference for the Willamette River. Gaining some additional insight into the river's history, issues, and current trends, coupled with an understanding of how it looks and feels, can provide much to consider.

Experiencing the river first-hand will improve your understanding of this important natural resource. There are numerous opportunities for anyone to get close to the river, whether you hike along it, sit at a riverside natural area, or travel by paddle craft. By traveling the Willamette River, you can learn the rhythm of the current, the color of the water, and the sounds of the swirling eddies on a long outside bend on a summer day. Seeing the rise of a great blue heron from a long fallen cottonwood snag or hearing the quick irritated staccato call of a kingfisher can help you develop a keener connection to the river and understand what creatures live there. It can also foster questions about the overall river system. What kind of habitat best supports birds like osprey or fish like lamprey? Getting close to the river can embolden and encourage you to learn more. That first sighting of a bald eagle on the river, perched high in a cottonwood, may hook you for the long term. It may even set you on a new pattern as you seek out evening canoe journeys to hold in view, however briefly, the deep red river sunset as the light disappears behind Marys Peak

A squirrel taken aboard at mid-river shudders in a canoe's bow. The little guy made it safely to land and the riverside trees.

west of Corvallis. As you gain experience with the river and begin to internalize some of what you've seen in a large greenway park or while fishing the riffles near Junction City, you may know that it is indeed time to give back to the Willamette in a meaningful way.

There was a time in the late 1960s and early 1970s when the Willamette River made national news for a great cleanup that took place. From Eugene to Portland, in many areas the river was highly polluted with raw sewage and toxic chemical wastes that were dumped directly in the river. Governor Tom McCall is the lone individual often associated with cleaning up the Willamette, but thousands of people were involved in this effort. Prior to his time as a politician, McCall served as a news reporter for KGW TV in Portland. In his special report "Pollution in Paradise," in one scene a dying fish was taken from the lower reaches of the Willamette, unable to breathe in the river's water. As a result of such coverage, along with the tireless work of many people, Oregon enacted laws to help clean up and better treat the waste that had been allowed all too freely into the Willamette and other waters around the state.

In his wonderful book, *Fire at Eden's Gate* (1994), Brent Walth described this part of Oregon's history and the importance of Governor McCall and Governor Bob Straub in providing leadership for improving the river's condition. Of course, their work was supported by many people and organizations, such as David Charlton and the Izaak Walton League among many others. Their work made a huge difference for the Willamette, especially with regard to its water quality. For all of the progress that was made, however, there is much left to do—and some things that absolutely need to be done.

Despite the tremendous efforts of earlier decades, the Willamette poses issues today for Oregonians and all those who care about rivers. We understand in a more comprehensive and meaningful way the essential components that a river needs to function properly. Much has been learned in recent decades about the relationship between the presence of clean cold water, healthy and abundant habitat throughout a watershed, and fish and wildlife. The understanding of toxic pollution has become more refined, leading to critical leaps in how we gather and interpret data from a host of species from the lowest to the highest on the food chain. Although many key concepts were well understood several decades ago, today we have even more information to help guide our actions for maintaining healthy rivers.

Working toward natural function in river systems is extremely important, and fisher-folk have long understood that when there is clear cold water in a river channel that is not highly restricted, fish can thrive. A naturally functioning river is one whose channel can meander to some degree from year to year and season to season, providing high-quality gravels for fish to rest or spawn in and off-channel areas for refuge. When these basics are interfered with, either by cutting down forests along streams or hampering stream or river flows, the consequence is easy to understand- -fish have a much tougher time thriving. In fact, populations of fish that were once commonplace in the Willamette are now decimated.

Paired with this earlier intuitive understanding of what rivers need, intensive research and data gathering have given us the knowledge to do what is right for the Willamette River. We need to continue to address toxic pollutants that affect the health of both wildlife and people. This became apparent in the 1960s when DDT and PCBs drastically reduced bird populations, seen most prominently in ospreys, bald eagles, peregrine falcons, and other birds of prey, and their subsequent resurgence when pollutants were removed. More importantly, researchers now understand that we must address environmental issues that merely have the *potential* to affect the river's ecological framework. A precautionary action taken today can do a lot to prevent significant harm down the road.

Although the Willamette River system has been altered dramatically in many ways, it can regain essential elements that make any river system healthy. The Willamette needs clean cold water and dynamic natural flows that aid habitat creation. Floodplains and side channels must better connect with the river or be reestablished altogether in some cases. Dams on the river need to provide natural flows and fish passage or be removed.

This newspaper box and garbage in the river attest to how some people just don't get it.

Pollution must be curbed from runoff from both urban and rural areas, and sewage pipes must be diverted from the river. Known contaminated areas, such as Portland Harbor, must be cleaned up. A tremendous effort is needed to provide widespread habitat restoration throughout the Willamette Basin, simply to give the river back some of what has been modified or taken over the past 150 years. We also must protect the river's higher tributaries and source waters, to have the best start at clean cold water that can provide greater opportunities for long-term river health.

Simple actions can make a big difference now and in the future. First, go out in your backyard to learn about the small creek that meanders through the neighborhood toward some Willamette tributary, for the environmental issues that affect creeks and larger rivers are much the same. Second, explore and enjoy the Willamette River. Local parks offer numerous opportunities to get close to the river. In many of these areas the flora and fauna are in surprisingly good condition, allowing us to recognize areas that need improvement.

Using this Guide

This book describes what the Willamette River is and how to experience it first-hand. It can also provide a good overall immersion in river topics that are timely and will be for some time, even if you cannot visit the parks or trips described in the book. Descriptions of how the river varies as it flows from south to north and how it behaves from season to season are provided. There is also ample discussion of its main issues, from water quality to threatened and endangered species to habitat restoration. While no single resource can fully capture the essence of an amazing river like the Willamette, this book can help you understand the river's history, its current status, and various trips and visits to the river.

We will follow the river by segments, with stretches of the river grouped thematically as it flows from south to north. There are no hard and fast border lines between the stretches, though, as what occurs in one area of the river may well be seen in other areas as well, harkening back to that mosaic concept. A good example of a recurring issue is that of riparian restoration. Loss of riverside vegetation is discussed in the early part of the book, but is applicable to the entire Willamette River system. The book begins at the southern end of the Willamette Basin, at the river's headwaters, and follows the river as it makes its way northward past dense forests, agricultural fields, and cities. In each stretch, you can learn about the area's key environmental issues, specific areas of interest for the river explorer, and how you can access them either by land or by boat.

The Willamette River system affords numerous opportunities to experience it from a paddle craft or by visiting one of the many parks that line the river and its tributaries.

For each area from the headwaters to the confluence with the Columbia, you will find suggestions for scenic vistas from land as well as for river trips that vary in length from 1 to 24 miles. The riverside visits, accessible by bike or car, and river trips for paddle craft or other conveyance offer the best opportunity to gain a better understanding of the river. The maps show river visit and river trip locations as well as natural areas and state and local parks.

Most sites include driving or river travel directions focused on river mile (RM) delineations. The RM examples reflect the commonly agreed upon mileage designations along the Willamette as used by the Willamette Water Trail, Oregon Parks and Recreation Department, and others. River mileage is somewhat counterintuitive along the Willamette. Although the river begins south of Eugene, the river mileage is calculated from the mouth of the river where it flows into the Columbia, making their confluence RM 0. As one travels upriver, the numbers increase to RM 187, where the Willamette begins.

Maps

The master map at the front of the book provides an overview of the entire Willamette River, from the headwater areas in the mountains to the tidal flats near the river's end. In addition, in each chapter there are detailed maps for the stretch described in the chapter. The trip headings and the maps give you the key details, such as put-in and take-out options, that can help you make your way. Included in these trips you will find information about what to see and to look for in each stretch. The maps also give a sense of riverside lands, such as floodplain areas and state and federal parks. However, these maps are not intended to be used for navigation on the river.

For those choosing to travel the river by boat, there are other more specific guides to consult with regard to river hazards and safety. The Willamette Water Trail Guide is a good set of comprehensive maps (available at www. willamettewatertrail.org). The Willamette Water Trail is an assemblage of public properties along the Willamette offering amenities such as river access and campsites. The trail enables one to travel the Willamette for a day or several days and is developed for paddle craft. The American Canoe Association also has an abundance of information on river safety, whether you travel by drift boat, canoe, or kayak. The Oregon State Marine Board has recently placed a heightened emphasis on boating safety and materials are provided on their website, although this agency is mostly attuned to powerboats. The Oregon Parks and Recreation Department also provides information about some of its major parks along the Willamette, with contact information for park professionals. See Resources for contact information and additional safety resources.

Canoe and Kayak Routes of Northwest Oregon (Jones 1997) has a good section on the Willamette. *Soggy Sneakers*, developed by the Willamette Kayak and Canoe Club (Giordano 2004), is a good periodically updated guide for running whitewater rivers, including some of the tributaries to the Willamette. These rivers can have far different conditions than the mainstem Willamette. A guide such as this should be consulted before canoe, kayak, drift boat, or raft trips on tributaries such as the Middle Fork, McKenzie, North or South Santiam, Molalla, and Clackamas Rivers.

River Safety: General Concepts

The goal of this guide is to enable you to learn about the Willamette River's environmental issues, history, and health, but there are a few things to keep in mind if you are traveling along the Willamette River. Regardless of the length of your trip to the river, either by car, bike, or boat, always bring water. Also, plan ahead and take a change of clothes, sunscreen, or other gear to help accommodate for changing conditions along the river. Imagine a freak rainstorm in June that lasts all day. For boaters, it is always a good idea to have a change of clothes, especially if you take an unplanned swim.

Rivers are dynamic. Although a map can provide excellent detail and point you in the right direction, conditions and routes can differ from one season to the next. A side channel that once provided clear passage around an island may have much reduced flows, making passage in any craft problematic. A channel may also be obstructed by woody debris, again forcing a change of approach. Thus, although a map can be of great value, trust your eyes over the map. If you are arriving at the river by bike or car, the same applies. It is up to you to be aware of your surroundings and the potential dangers that may exist. The river channel and conditions at greenway parks and other natural areas can change over time. Always travel within your ability.

No matter the situation, traveling any river by boat presents risks. For instance, in the Upper Willamette, where the McKenzie enters the river near Eugene, the river has many gravel bars and an ever-changing channel. This can be hazardous to powerboats, especially ones with propellers. Many of the suggested trips in the book assume that you will travel by paddle craft, which can be a less intrusive way to access the river's quiet areas and likely a better opportunity to view wildlife. The following are a few critical things to consider when traveling any river.

▶ You are responsible for your own safety. Never assume that any guide or book can provide all of the answers or keep up to date with changing river conditions. Always be on the lookout for potential hazards.
▶ Understand your level of skill in relation to the river that you are traveling. If you do not have recent experience or have not had recent paddling

instruction, it would be unwise to assume that you are ready for the challenge that a river like the Willamette, or any other, can present.

► Always wear a personal flotation device (also known as a life jacket). This is a simple thing to remember but is far too often neglected. You may be a strong swimmer and feel very confident about your ability, but the truth is that when people get injured or die on rivers, the most frequent cause is not wearing a life jacket. The Willamette has a strong current in many areas, a force that can make it extremely difficult to swim to shore if you flip your kayak or canoe. Always wear a personal flotation device.

► Check river flows, especially in winter and spring.

► Many Willamette tributaries have stretches of rapids. Again, do not travel sections with rapids by paddle craft or powerboat unless you understand how to negotiate rapids in these craft and understand the level of danger in these areas.

► Strainers—woody debris such as tree root wads or limbs that are stationary in the current—can be deadly. At all costs do not make contact with strainers, either with your boat or your body.

► Did you catch the mention of wearing your personal flotation device?

Experienced or not, one should always be prepared for the unexpected, and to help others in need. These very skilled paddlers are very relieved after a close call. Duct tape can sometimes help a trip continue.

When traveling the Willamette River or riverside areas, it is a good idea to know how the river is flowing. There is a big difference between late summer flows, which typically have a relatively small amount of water, and winter and spring flows, which can have huge volumes of water and even occasional flooding. Over the years I have heard numerous stories of friends who decide to take a "nice little kayak trip" along the river at higher flows. If you have not traveled the river at high flows, you have not experienced the swirling currents and eddy lines that can test your ability to stay relaxed and paddle. High flows can affect land access to some areas as well, making parking areas inaccessible or hikes into riverside areas unrealistic. Hourly river flow updates can be viewed at the U.S. Geological Survey Website (http://waterdata.usgs.gov/or/nwis/rt) and at local or regional sites.

Natural History

One of the best things about traveling the Willamette River is that you have a good chance to experience the natural world up close—not only in remote greenway parks, but even in urban areas. This guide provides some information about the Willamette system's geology and common plant species, as well as abundant information about fishes, mammals, and birds. Common native species along the river's entire length include cutthroat trout, spring chinook, great blue heron, green heron, osprey, bald eagle, deer, river otter, beaver, and more.

Take the time to experience and enjoy the Willamette through the many parks that exist up and down its reach. Some argue that we have lost a critical "river ethic," or perhaps that a river ethic has never been fully incorporated into our daily lives. Such an ethic, which places value on maintaining clean cold water, healthy and abundant habitats that support a range of native species, and open public spaces, is not so farfetched. In time, if you and others like you engage with the Willamette and its tributaries, a new river ethic can well take hold, making clean and healthy rivers commonplace throughout our region.

Useful River Travel Terminology

RIVER RIGHT: while facing downstream, the portion of river and bank to the right

RIVER LEFT: while facing downstream, the portion of river and bank to the left

STRAINER OR SNAG: typically a tree trunk, branch, or roots that can stop you and your boat and hold you in place. Strainers can be in the middle of the river and appear as a branch bobbing in the current, a tree or collection of roots sticking into the water from the riverside, or any combination in between. Strainers can be deadly and should be avoided at all costs.

The Name of the River

While there is no certainty, according to *Oregon Geographic Names* the Willamette River likely gained its present-day name from the Kalapuyan peoples, from the word Wal-lamt. Over the years several spellings have been recorded, from Wilarmet and Wallamette to Wallamat and Wallamatte. This or a similar word likely referred either to the river at Oregon City or to the stretch of river above Willamette Falls. The word may mean "green river" or "spill water" according to early accounts. In 1792 William Broughton named it River Mannings, and as early as 1838 the U.S. government referred to the Willamette in its lower reaches as the Multnomah River. In 1841 Charles Wilkes used the present spelling, and it stuck.

Where the Willamette Comes From

Men may dam it and say that they have made a lake,
but it will still be a river. It will keep its nature and bide its time.
Like a caged animal alert for the slightest opening.
In time it will have its way; the dam, like the ancient cliffs,
will be carried away piecemeal in the currents.
—WENDELL BERRY

YOU MAY WONDER, what is *the* source of the Willamette River? Like most rivers, the Willamette has no single source that creates it. Instead, the headwaters are a collection of springs, creeks, streams, and thirteen major river tributaries that form the river system. Because the river flows from south to north, however, the key source tributaries are in the southern part of the nearly 12,000 square mile Willamette Basin.

The Coast Fork and Middle Fork form the mainstem Willamette just south of Eugene. In some ways, the Middle Fork is actually the mainstem, given its character and the significantly greater volume of flow than that in the Coast Fork. The McKenzie River flows in just 12 miles downstream from the start of the mainstem river. Because the flow from the McKenzie adds substantially to the Willamette's volume, we can look to the high-mountain headwaters of both the Middle Fork and the McKenzie River as some of the key source waters of the Willamette.

The Middle Fork system holds perhaps the crowning element of the Willamette's source waters, Waldo Lake, located about 90 miles southeast of Eugene. High in the Cascades within the Willamette National Forest, the lake sits like a massive jewel cupped in the heart of a vast conifer forest. Nearby and to the south is Diamond Peak, with gleaming snow-capped heights that can be seen well into early summer. Waldo Lake sits at 5414 feet in elevation and is approximately 8 miles long and more than 400 feet deep. With some of the purest of water, the lake provides a contrast to what can be found further down the Willamette Valley, where the waters are less than pristine. At the northeast end of Waldo Lake, a small stream of water pushes northward, starting the North Fork of the Middle Fork of the Willamette River. This surge of water is small, but very significant. Waldo Lake has no tributaries that flow

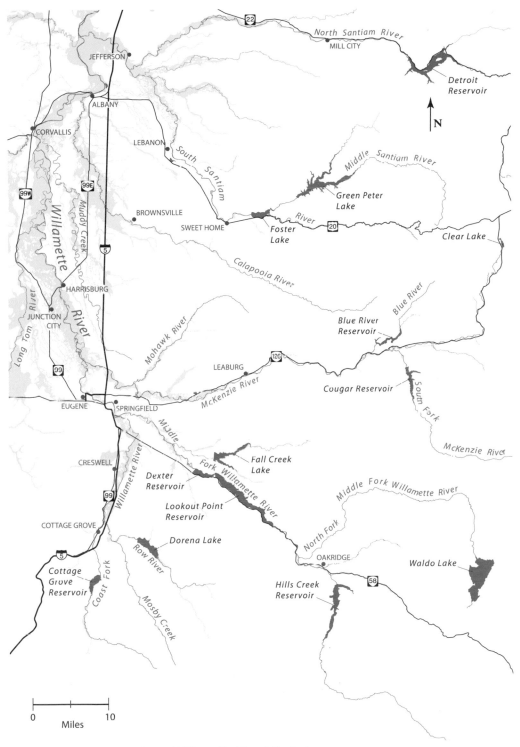

Upper Willamette River tributaries

into it, but is instead fed by groundwater, snowmelt, and rain that slowly charge the lake over time.

Like much of the rest of the Willamette River system, Waldo Lake has seen its share of human impacts. In the early 1900s, Waldo was the center of plans to deliver irrigation water to downstream areas via a massive tunnel that was to feed water into Black Creek. The irrigation company speculators assumed that snowmelt recharged Waldo with water each winter and spring. The Waldo Lake Irrigation and Power Company, an enterprise established by Amos Black, Frederick Ray, and Simon Klovdahl, claimed water rights on the lake and began preparation for the massive diversion at its northwest end. Although the venture was hampered by a lack of funds, they blasted a tunnel several hundred feet into the mountain. The company eventually failed in 1934, leaving behind a tunnel that leaked a massive amount of water from Waldo Lake. The tunnel was sealed with concrete in 1987, but the remains of this structure can still be seen at the lake today.

While one may marvel at the idea of taking water from and depleting such a pristine lake, the story reflects what has happened throughout the Willamette River system. Like many other areas in the Willamette Basin, in the early days the need for resources was often met with little thought of the environmental consequences. Waldo Lake has stood the test of time and is a fitting headwater for the wondrous Willamette River. Protecting this pure water source is a noble yet necessary goal for the public.

Waldo Lake on a cold autumn afternoon

In 2007 the management of Waldo Lake by the U.S. Forest Service entered a somewhat controversial phase. The agency sought to minimize human impacts to the lake's water and aesthetic quality by phasing out crafts with gas-burning motors. This effort aimed to prevent pollution from spoiling the lake's waters and keeping noise at the lake to minimum. While the overwhelming majority of the interested public has supported the Forest Service's proposal, the plan was appealed by a small group of individuals. These few seem to believe that no place should be off limits to the noise and pollution of boat motors. This issue continues to be addressed at the time of this writing. If you choose to visit Waldo Lake, however, you may also come to believe that some places are simply too special to take a chance of being spoiled.

Waldo Lake Area

WHERE: Heading southeast from Eugene, take Highway 58 approximately 18 miles south of Oakridge. You will see a clearly marked sign for Waldo Lake, up Forest Service Road 5897. There are several places to stop to view the lake, although the best may be from the North Waldo Campground.
AMENITIES: There is parking at the campgrounds around the lake (Forest Service day use pass required). If you choose to hike, bring water.

Camping is available at North Waldo in the developed camp area. For a more rustic experience, you can paddle around the lake's perimeter to one of several areas that make great campsites. Crossbills, mergansers, and other birds may visit your camp. You can also paddle to the small slip of water that creates the North Fork of the Middle Fork of the Willamette at the northeast finger of the lake. The full perimeter of Waldo Lake is 21 miles, with a width of between 1 and 3 miles. It is a good idea to stay near the shoreline, paddling around the perimeter to get from one side to the other. The wind can rise quickly, at times generating significant whitecaps and waves.

The pristine nature of Waldo Lake is easy to comprehend by simply looking down into the water at the lake's edge or from a boat. On calm days, you can easily see the lake's white sandy bottom. The clarity of the lake's water is fantastic, enabling you to see well over 50 feet down.

Autumn is an excellent time to visit, as the weather can be fantastic and the mosquitoes are well past their peak. An early autumn morning with the deep orange cast of the sunrise can be magnificent. Waldo is also valued by many who travel there for the quiet and relative solitude. It is a place where you can easily be immersed in the peace of the water and the dense forest surrounding the lake. ■

Waldo Lake is a striking image of perhaps what a headwater should be, a vast lake of pure water feeding a small mountain stream. However, another type of headwater area is just as striking: the McKenzie River watershed. High in the Cascades the snow melts away in early summer, feeding a host of creeks that eventually flow into the McKenzie River's clear cold water. Over the summer months, the snowmelt high in the watershed is expended, and the creeks that raged only weeks before can taper off drastically, leaving many of these small waterways mostly dry. This is true of many seasonal creeks in mountain ranges across the West. Yet high in the McKenzie watershed, water rushes down mountainsides year-round in places, feeding the main river. These unusual sources of water are headwater springs.

Much of the High Cascades is basalt, a common porous rock that is associated with volcanic eruptions. The vast field of volcanic rock that rests at the top of the mountain range acts like a massive sponge, soaking up and retaining the snowmelt as groundwater. The High Cascades are geologically newer than adjacent areas, with basalt that is generally more conducive to groundwater flow than other geologic formations. Over time, the snowmelt and rainfall percolates downward through the spaces between the rocks and the pores in the rock itself. Eventually the rivulets of subterranean flow join others, forming a vast aquifer deep within the Cascades. This massive natural reservoir spreads out across a vast area. With such a high volume of water deep in the earth, natural pressure points are created, thus enabling surging water to burst forth out of the rock itself. These are the springs in the McKenzie system. While this phenomenon also occurs to some degree in the Santiam and Molalla River systems, two other Willamette tributaries with headwaters derived from the High Cascades, this trait is most pronounced in the McKenzie drainage.

Olallie Spring is one of the most dramatic springs that feeds the McKenzie River. High up in the Cascades and few miles from Highway 126, Olallie Spring is in the heart of the forest. The area has no signs and no parking, and it is off of a relatively nondescript Forest Service road. After walking down a steep trail for some time, you'll begin a sudden transition from the quiet of the forest to the sound of rushing whitewater, as if there is a waterfall somewhere ahead. The temperature drops noticeably, with cool moist air filtering through the area. As the slope unfolds, suddenly the view opens up and the smell of water spreads outward. A few more steps and your senses are overwhelmed by a blast of whitewater streaking forth from a hillside surrounded by green moss and ferns, forming a vast torrent that enters Olallie Creek below. The water of the spring is ice cold as it surges from the hillside, having moved miles underground through an endless sea of basalt, starting somewhere in the wind- and ice-swept plain high above. Drawn by the gravity of the earth through vast networks of dark porous rock, the water finally finds its release at the opening in the hillside near Olallie Creek. It is an amazing sight to behold. Over the years, several springs have been found, mostly in extremely remote locations.

Unlike Olallie Spring, nearby Great Spring has relatively good access and is adjacent to Clear Lake. At Great Spring a deep pool forms where the water streams out of the basalt bank near an overlooking trail. The rich greenish blue water is very cold, serving as an excellent source for the river system. Just a couple hundred feet downstream, the flow visibly pushes its way into Clear Lake, which was formed about 3000 years ago when the river was naturally dammed by a geologic upheaval.

Springs like Olallie and Great Spring are critical to sustaining the overall volume of water flow in the McKenzie River. In recent years, our understanding of the influence of these springs on the McKenzie and other river systems has been advanced by the work of Gordon Grant and his colleagues at the U.S. Forest Service in Corvallis. A network of these high mountain springs provides a significant portion of the water volume to the Willamette River, via the McKenzie River, during the low-flow months of summer. The flow from the springs is generally quite cold and amazingly consistent year-round. Given the amount of flow that is diverted for various uses, the Willamette might be starved without this water. Roaring Spring, the largest of the springs identified to date, provides a meaningful amount of flow to the mainstem Willamette in Portland Harbor in the heat of the summer. This is an amazing fact, given that the harbor is more than 200 miles downstream from the spring.

Great Spring at Clear Lake

WHERE: From Springfield, take Highway 126 east to Clear Lake.
AMENITIES: Restroom and parking lot at Clear Lake.

You can take a row boat, which can be rented at nearby Clear Lake Resort, to the north end of the lake and catch the trail that travels the lake's edge. A short walk will take you above the rich blue pool that is Great Spring, which is marked with an old wood sign on the tree. Take a moment to touch the spring, but get ready for some very icy water. This large pool gets its water from the basalt at the water's edge. It then flows a couple hundred feet down to Clear Lake. This is an impressive opportunity to see the source of some of the McKenzie River's water and to learn how other springs in the area contribute to the Willamette's flow. ■

Source waters of other tributaries are also critical to the Willamette system's health. Although several other Willamette tributaries have important headwaters as well, the southernmost tributaries are the most defined sources. In addition to the Middle Fork and McKenzie River, the small streams that feed the Row River and the Coast Fork, nestled deep in the Coast Range, can be included in that group.

For a time people did not acknowledge the importance of high mountain streams to the health of rivers far below and the myriad species that rely on healthy tributaries, such as trout and the animals that feed on them. In more recent times the linkage between the health of heavily forested areas in the Cascade and Coast Ranges has been better understood. But too often small creeks have been hammered by logging, which can eliminate shade that keeps the creeks cool and can cause erosion that carries sediment into creeks, thus disrupting the food cycle in these areas. Although today logging practices have improved to some degree, there are still logging companies who seem to work against common sense. In the name of getting a few more trees out of a forest, proposals have gone forward to vastly decrease the amount of riparian buffers and to thwart other practices that help to protect headwater areas. If the desired outcome is to have healthy rivers in the valley bottoms, we need to protect areas in the upper parts of a watershed as well. Ideally, all the sources of the Willamette's tributaries should be well cared for, because protecting clean cold water high in the mountains benefits the larger rivers far downstream.

The Major Tributaries

Cold water rushes downhill, spreading among boulders and smooth rounded river rock. To the side is a stand of Douglas fir and at the riverside cottonwoods fill in the lower reaches. The water dances in pulses, some tamed by dams high above, and other flows making their way to the river naturally, each driving the pulse of the current downstream. Deer intermingle in the riparian fringe, peeking out to see a small, silent boat pass by. Mist holds at the river's surface, with a stray ray of sunlight cutting toward the water. Lively, pushing, and ever moving, the three major rivers at the start of the Willamette have a similar character. The Coast Fork, Middle Fork, and McKenzie help define the larger river downstream.

Coast Fork of the Willamette

Entering the Middle Fork from the southwest, the Coast Fork of the Willamette is a little different than the two others. It derives its flow from both the Calapooya Range in the Cascades and from the Coast Range, and overall it has a smaller flow than the Middle Fork and McKenzie River. It adds the Row

River, a key tributary of the Coast Fork, just downstream of Cottage Grove. In winter and early spring, this tributary can provide very high flows, as it catches a large volume of water from the rain-drenched Coast Range southwest of Cottage Grove.

The Coast Fork extends approximately 55 miles from the headwaters to its confluence with the Middle Fork of the Willamette. These two rivers have similar characteristics, though they reflect watersheds of different sizes and levels of flow. In recent times the Coast Fork has been little traveled. Like much of the mainstem Willamette, this nearly 600-square-mile river basin is highly impacted by hydropower dams, the largest being the Cottage Grove Dam. The Coast Fork is also affected by agriculture along its banks. Over the years a sizable amount of farmland has come very near its shoreline, with the most dynamic areas having been riprapped (lined with rocks) to keep high flows from reaching the historic floodplain.

Meandering along the valley bottom, the Coast Fork runs about 25 miles from Cottage Grove Dam to its confluence with the Willamette River. The river has few access points between the dam and Cottage Grove and little in the way of parkland along its reach until it passes Creswell. Below Creswell the Coast Fork has three sizable greenway parks: Camas Swale, Bristow Landing, and the wonderful county-run Mount Pisgah area. As with other areas in the Willamette system, greenway properties along the Coast Fork provide a view into the river's past, however fragmented that view may sometimes be.

Along one of the river's bends, Camas Swale between RM 9 and 10 provides a wide plain of rounded river rock, dotted with willows expanding outward from the shoreline. In warm months, the area is bustling with the chirping of sparrows, warblers, and other birds. The river arcs to the right at Camas Swale, creating a relatively sharp bend that forms a rounded peninsula of river rock. About 100 yards to the other side, the river completes its bend over a low gravel shelf. At its base, this peninsula plays host to massive old black cottonwoods (*Populus balsamifera* ssp. *trichocarpa*), with deer hidden in the deep foliage underneath and the sounds of the wind riffling through the leaves atop the old trees. This large greenway park is undeveloped. Camas Swale receives little use, because it is inaccessible by land on a public road and is situated along a small river that few people travel.

Just upriver from Camas Swale, Bristow Landing provides much the same, with a large public area accessible by river only. Here an abundance of bird and mammal life can be seen, including osprey, beaver, and belted kingfisher. Downriver from Bristow Landing reality sets back in, as houses begin to appear along the bank. Not far ahead is Highway 58, with the bridge crossing and a small boat ramp underneath. Just downstream, though, is a real treat for those seeking a large natural area along the Willamette system, the Howard Buford Recreation Area. As you round a bend and head back north, you'll notice a change in the geology, the appearance of rock at the river's surface. This small shelf of rock and the rise of the land on river right indicate the boundary of the Buford Recreation Area.

In 1970 the state of Oregon purchased 2300 acres of land at the confluence of the Middle Fork and Coast Fork Rivers, creating the Mount Pisgah State Park. The state leased the acreage to Lane County in 1973, when it was renamed Howard Buford Recreation Area. The dominant landform in the recreation area is Mount Pisgah. This large natural area is on the southern boundary of the Eugene metro area. The park and its adjacent arboretum is an amazing asset, with an open natural area, hiking trails, and the wonderful summit of Mount Pisgah reaching a height of 1531 feet, some 1000 feet above the valley floor. Mount Pisgah makes a fitting overlook for the official beginning of the Willamette River, affording views well into the Middle Fork and Coast Fork Watersheds. While a paddle craft cannot easily be put in at the Buford Recreation Area, it is well worth a stop. There are a few areas just upriver of the bridge crossing into the park and adjacent to the paved parking area that can enable disembarking from a canoe or kayak. Better yet, traveling by car or bicycle will allow you to enjoy several hours traversing the hillsides.

Since 2006 the Friends of Buford Park and Mount Pisgah (see Resources) have worked with many others to acquire land owned by the Wildish Sand and Gravel Company. This company filed a Measure 37 claim on their land adjacent to the Middle Fork and the Buford Recreation Area. The claim asserted that local zoning made it impossible to develop homes along the Middle Fork. Given its proximity to the beginning of the Willamette and the recreation area, the filing of the claim was met with much concern. Many have encouraged the state to purchase the land owned by Wildish, adding to the park, and momentum is growing for this outcome. Having more of the confluence area in public ownership would provide a good start for the mainstem Willamette with a large natural area replete with native vegetation and a wonderful level of commitment by local interests into its health and longevity.

Howard Buford Recreation Area

WHERE: Southeast of the Eugene/Springfield area. The park is east of Interstate 5 and north of Highway 58. Take the 30th Avenue exit off Interstate 5, and head east on Franklin Boulevard to Seavey Loop Road. After 5 miles, this winding road crosses a bridge over the Coast Fork and soon leads to the main entrance and the North and Main Trailheads. The route is well signed.
AMENITIES: Fee parking, a visitors' center, restrooms, and running water. Just off the main parking area is the arboretum, a separate but related operation of some 200 acres that grows native Northwest plants.

The park is along the Coast Fork and Middle Fork, and the open grassy hillside is dotted with Oregon white oaks (*Quercus garryana*). For an excellent look at the confluence that begins the Willamette, hike up the northwest side of Mount Pisgah. As you gain elevation, views of the confluence unfold. You can also get a good sense of the basins drained by the two tributaries, especially from the summit. Expect a good 45-minute walk to the top. Standard day-hiking equipment works well for this site, and bring drinking water. ■

The Coast Fork affords a close-up view of the manipulation of a river system by people. Along the Coast Fork, human intervention is clear, with a flow controlled by long stretches of revetments. A revetment is a human-built structure put in place to help stabilize the river's bank. Most commonly revetments are composed of large rocks or riprap placed at the water's edge and on the bank, sometimes reinforced by wood pilings. These structures create a high bank that is kept in place and resists the river's need to reach into lowland riverside areas. In addition, like many other Willamette River tributaries, the Coast Fork has two hydropower dams that regulate the river flow and are run by the U.S. Army Corps of Engineers. The Cottage Grove Dam controls the flow on the Coast Fork year-round, as does the Dorena Dam on the Row River.

This dam has altered the natural hydrograph of the Coast Fork. The hydrograph of a river is a numerical representation of a river's flow over the course of a year. When a river's hydrograph is natural, it flows according to how much rain or snowmelt is deposited in the river basin and the subsequent

An early and dirty effort to stabilize the bank of the Willamette in Polk County, 1939. Oregon Historical Society, no. OrHi55639

movement of the surface water and groundwater that flows into the river. Flows will be large in the spring and then taper off until summer, when flows are low. Most rivers with a natural hydrograph provide amazing surges of flow, with water levels that can rise very quickly. This was true of the Willamette River until the 1930s, just as it is true today of the Molalla River, one of the Willamette's west-side tributaries. In the case of the Coast Fork, the hydrograph is manipulated by the Cottage Grove Dam as well as the Dorena Dam on the Row River, a major Coast Fork tributary, which controls how much and when water is released. The hydrograph can vary within a given season more than once.

As with other rivers with dams, the Coast Fork has lower than normal flows in the winter and spring and slightly higher than normal flows during summer months. Before the dams, high spring flows helped to carve the valley bottomlands, with the river depositing sediments and developing new side channels that replenished backwater areas with sediment and nutrients. This is only one significant way that dams have changed the Coast Fork, and such changes occur in all dammed rivers.

Coast Fork, Cloverdale Access to Mount Pisgah, RM 12.5 to 2.5

STARTING POINT: From Interstate 5 take Exit 182 and head east on the Springfield-Creswell Highway. After crossing the bridge, take the first right and you will see the entrance to Cloverdale Access, an Oregon Parks and Recreation Department property.

ENDING POINT: Buford Recreation Area, located southeast of Eugene, just east of Interstate 5 and north of Highway 58. You can access the entrance off Seavey Loop Road by taking the 30th Avenue exit from Interstate 5, then heading east on Franklin Boulevard to Seavey Loop Road. The route is well signed.

DISTANCE: 10 miles

SKILL LEVEL: Moderate paddling skills.

CONDITIONS AND EQUIPMENT: This section of river has ample current, though summer flows can be quite low, and occasional snags and rocks. It is suitable for a canoes and kayaks year-round.

AMENITIES: There is parking at both Cloverdale Access (free) and at Mount Pisgah (fee required). Cloverdale has portable toilets.

WHY THIS TRIP? Scenic views of a lively river that is not often traveled

Situated on the east side of the Coast Fork, the Cloverdale Access park is a bit out of the way. But this put-in along the river is the start to a very enjoyable, and likely surprising, interlude with this Willamette tributary. Just downstream of the park, the Coast Fork offers up a small rapid. While the flow of the Coast Fork is tightly controlled by the U.S. Army Corps dam upstream, the river seems to have a lively energy even in low October flows.

From a canoe traveling through the shallow but constant current at Bristow Landing and Camas Swale, the vision is one of inviting green. Cottonwoods tower behind the willows and other trees that dot the shoreline. For almost a mile of beaver chew sticks and small depressions in the bank, the Coast Fork brings the river traveler through the floodplain and richly forested habitat on these state-managed properties. Fronted by quick-moving water and occasional small standing waves, Bristow Landing continues on river right for nearly a mile, ending at a shallow backwater. Just above the backwater area is a weathered sign that simply states "10.6," which at first glance might be mistaken for 106. You can paddle just past the backwater and enter a large shallow wetland area immediately adjacent to the river, where there is plenty of area to explore.

Soon you'll see more river rocks at a left turn, as Mount Pisgah comes into view. You can take out at the Mount Pisgah Landing. The Coast Fork in this stretch is small, with occasional riffles. The confluence area is not dramatic, and the Middle Fork can be seen coming in to the right. There is a small Oregon Parks and Recreation Department park on an island that affords a good view right at the confluence. ▣

Although it plays a critical role in forming the mainstem of the Willamette, the Coast Fork is a river that is often passed over, sitting in the shadow of the Middle Fork and McKenzie. Yet if you take a little time to get to know this blue ribbon flowing from the Coast Range, you will likely develop a lasting appreciation for it. For most of its 55-mile length, the Coast Fork is relatively small. Paralleling Interstate 5 and Highway 99 for a good distance, the river is dotted by the communities of Cottage Grove and Creswell. As with so many of the basin's rivers, one need only take a short drive to arrive at a secluded piece of land along the Coast Fork of the Willamette.

Greenway parcels or parks are large undeveloped tracts of land adjacent to the river (unless the river has moved—a topic to be covered in later chapters) that wildlife and people can use, such as Bristow Landing and Camas Swale. The Willamette Greenway Program was founded back in the early 1970s. The most significant and most visible elements of the Willamette Greenway Program are the parks managed by the Oregon Parks and Recreation Department, which owns and manages multiple properties along the Willamette River and its tributaries. The parks can vary from secluded spots where few people tread to large areas that require paid parking, with a host of facilities. While there are state-run parks along some of the tributaries, such as the

Coast Fork, the vast majority of greenway parks are along the Willamette mainstem. These are typically undeveloped blocks of riverside land.

Some greenway parks along the river have many facilities, including running water, parking, restrooms, picnic areas, and other amenities. While these parks are prized in many ways by valley travelers, most of the greenway parks are on the other end of the spectrum—with few facilities, no development, and little in the way of access by land. In fact, these more rustic assemblages of riverside lands offer the greatest level of scenic value, with a variety of wildlife and riverside forest areas and relative quiet. These undeveloped areas are typically the most desirable for those who go birding or fish and certainly for those who travel the river by paddle craft.

Originally, the greenway idea was based on developing a continuous assemblage of parkland along the mainstem Willamette, providing a vast public resource. To move this concept forward, in 1967 matching grants were given to local governments by the state of Oregon to acquire land along the Willamette River. Because of the inability of local governments to match the funds provided by the state, few land purchases were made and the program soon reverted back to the state. Not long afterward, the Oregon Department of Transportation, which at the time oversaw the Oregon Parks and Recreation Department, took the lead to keep the greenway idea alive and sought to purchase land along the river. Unfortunately, as part of this land acquisition effort, action was taken by the state to condemn certain properties. This led to a significant uproar among the citizenry, in time depleting the critical energy needed to complete the greenway system. As you might imagine, there are few better ways to foster a lifelong grudge than to condemn and take away a person's land.

In 1973 the state legislature passed the Willamette River Greenway Act, which required the creation of a Willamette River Greenway Plan that had to be adopted by the Department of Transportation and the newly formed Land Conservation and Development Commission. Unfortunately, the plan was not approved by the commission. While this effort hit another hurdle, there was still significant interest in the greenway system among the legislators, Governor Robert Straub, and the general public.

The Land Conservation and Development Commission then created Land-Use Planning Goal 15 to replace the first plan. Goal 15 directed the Department of Transportation to create a plan that provided proposed greenway boundaries, access points both existing and proposed, and proposed acquisitions. This greenway plan also required significant input from the public, which was lacking in the original attempt. The original language sought to protect riverside lands, with the mission of Goal 15 being: "To protect, conserve, enhance, and maintain the natural, scenic, historical, agricultural, economic, and recreational qualities of lands along the Willamette River as the

Willamette River Greenway." More specifically, in Section A1, the authorizing legislation stated, "The qualities of the Willamette River Greenway shall be protected, conserved, enhanced, and maintained consistent with the lawful uses present as of December 6, 1975."

Over the years, many people have claimed that the date specified led some landowners to cut down trees along the river prior to that date, as they believed the legislation would prohibit such actions afterward. Unfortunately, as a result some riverside forest areas and buffers were negatively impacted. Goal 15 also placed some restrictions on riverside development along the Willamette River. These regulations had to be incorporated by local governments into the comprehensive plans and related codes, limiting how close building can occur to the river. In reality, Goal 15 has been applied unevenly among counties and cities.

Today the Greenway Program that exists is not what was originally envisioned. Too much opposition to the original effort hampered the ultimate strength and success of the program. Indeed, mention of the Willamette Greenway Program with some people is met with a roll of the eyes and comments about the program's relative lack of results. The absence of a consistent greenway along the Willamette is easily verified by a flight above the river. In many places there are no trees at all next to the river, and in many others only a very thin strip of riparian vegetation exists next to the river, often connected with heavily riprapped banks. Although the ultimate goal of a continuous band of protected land along the river has not been met, much progress has been made in acquiring public lands along the river. Today many of these parks provide some of the best places to visit the Willamette.

The greenway parks are generally quite scenic, offering numerous opportunities to view wildlife along the river. In many places you can paddle a canoe up to a large site along the river, with a small gravel bar or flat grass and sand mix suitable for a tent, backed by a mix of willows and cottonwoods. Many of these areas provide wonderful opportunities for solitude. Some are large enough to provide a bit of space to explore and wander along the river or one of its backchannels. In such areas, you might hear the flowing water punctuated by the slap of a beaver's tail or an owl hooting somewhere far behind the riparian fringe.

Since 1973 the Oregon Department of Transportation and more recently the Oregon Parks and Recreation Department have acquired more than 4000 acres of land along the Willamette River. In 2006 the approach to the Greenway Program was revisited and the Oregon Parks and Recreation Department developed a new plan that will more vigorously target riverside land for acquisition. The department has increased funding and staff who will work with willing landowners to acquire more properties in the coming years.

In addition to providing areas where people can recreate, greenway prop-

erties also have a great potential to provide significant benefits for habitat restoration and protection. Everything from increasing connections to floodplains to riparian planting could occur on many of the existing public parcels,
as well as future acquisitions. Today this habitat aspect of the Greenway Program remains the least documented and perhaps least understood by the
general public. In future years the Willamette River Greenway Program may
come closer to meeting its original intent. One can only hope that this is the
case, for as more people move into the Willamette Valley, there will be fewer
and more expensive opportunities to purchase riverside lands.

Middle Fork

The Middle Fork is often considered to be "the Willamette," as it appears to be
the most logical extension of the mainstem. Those who named the river, however, decided that the point at which the Coast Fork enters the Middle Fork
was the start of the mainstem Willamette. The Middle Fork watershed of 1355
square miles encompasses the communities of Oakridge, Westfir, Lowell,
Dexter, Fall Creek, and Jasper, among others. It also includes the fantastic
North Fork of the Middle Fork and the pristine Waldo Lake headwaters. The
Bureau of Land Management and U.S. Forest Service own much of the upper
watershed, and historically logging has been a significant issue in this area as
it relates to stream health.

Waldo Lake holds some of the cleanest and most pristine water anywhere.
The relatively small amount of water leaving the northeast edge of the lake
supplies the North Fork, which joins the Middle Fork near Oakridge. Between
Oakridge and Dexter, the river is lively. Although Highway 58 and other roads
come near the river at many points, the Middle Fork makes its way past some
alluring landscape.

Downstream from the Dexter Dam, about a quarter mile from the primitive boat ramp, the right channel around the first public island provides an
almost immediate sense of solitude. Here the Middle Fork emits a peaceful
essence, where the morning mist can mix with the smell of autumn leaves
dropping into the fast current. Although the entire flow of this key tributary
is controlled by the dams upstream, autumn rains can pulse the Middle Fork
Watershed with new life after the long, warm summer.

Four major U.S. Army Corps of Engineers projects are found in the Middle
Fork between Dexter and Springfield—the Dexter, Lookout Point, Fall Creek,
and Hills Creek Dams—with each section of waterway held back by its respective dam. The flow of the Middle Fork is tightly controlled to meet the
flow requirements of the mainstem Willamette River. In the normally dry
months of summer and early autumn, the Middle Fork maintains a relatively
high flow—to the top of its banks at times. Under a natural hydrograph, this

river would have greatly decreased flows during this time, but Corps dams and reservoirs feed the Middle Fork water during the dry months.

While this section of river is pushed to meet the mainstem's flow requirements, it retains a sense of wildness, with a multitude of channels below Dexter. Cottonwoods intermingle with fir trees all along the Middle Fork, and below the Dexter Dam the river meanders to and fro, divided by lush islands creating side channels of varying widths. Alcoves dot the extent of the Middle Fork, providing refuge for the threatened bull trout just beyond the fast flow of the main channel. Alcoves are fingers of slow-moving water that extend back from the main channel, dead-ending in shallow waters and wetland areas. They are important features of the entire Willamette River system, providing areas where larger fish can migrate to spend the night, rest, and gain shelter from predators. Juvenile chinook may also use these areas on their way downriver to the Pacific. On the Middle Fork, in particular, alcoves have been identified as areas where the threatened bull trout can be found.

Like other Willamette tributaries, the Middle Fork offers up whitewater ranging from more challenging runs above the dams to the occasional Class II rapid in the 16-mile stretch below Dexter Dam. The breadth of public land along the Middle Fork is exemplified by the properties just below Dexter, with Dexter Landing on the west side of the dam transitioning to the scenic Elijah Bristow State Park. These two properties create a long complex of riverside forests and former floodplain that provides an opportunity to restore some

Bull Trout

Bull trout (*Salvelinus confluentus*) was officially listed as a threatened species in 1998. These fish were once common throughout the Pacific Northwest, but today the populations in Oregon are at great risk of extinction. Bull trout have pale yellow spots along the back and red or orange spots along the sides. In general, they have light spots on a darker background, as is the case with all of the chars—true trout have dark spots on a lighter background. The leading edge of the fins is white and the dorsal fin is translucent. It is illegal to fish for bull trout, and those caught while fishing should be released immediately.

Bull trout need clean cold rivers with complex habitat, including a variety of pools, riffles, and water depths, and clean gravel with upwelling ground water is critical for spawning. As protection from predators, bull trout also need undercut banks, overhanging vegetation, and woody material. Logging activity may destroy stream habitat structure and overhead vegetative cover, which causes higher water temperatures and sedimentation of spawning gravels. Because of their dependence on high-quality habitat, bull trout can serve as a biological indicator of stream health.

floodplain function. In this case, floodplain function is synonymous with more natural (higher) flows being released by the dam.

While the destruction of the 1996 floods was unwelcome, flooding is a natural occurrence in river systems. The native fish species that inhabit the Willamette and its tributaries were long accustomed to seasonal flooding. When flooded, low-lying areas lining a river provide areas with slow-moving water for rest and rearing. Some of the thirty-one fish species native to the Willamette system utilized floodplain areas at key parts of the year, either for

Whitewater Rapid Classification System

When someone goes rafting or kayaking on whitewater rapids, it is vital that they have a good idea of what they can expect from the river. To simplify this process, all whitewater rapids are rated on a scale of I to VI. The rapids receive ratings based on a combination of difficulty and danger, where a Class I rapid is the least difficult and dangerous and Class VI rapids are the most difficult and dangerous.

The classification system for whitewater rivers is not an exact science and it may vary with fluctuating water levels. Typically, high water levels increase the difficulty of rapids, but this is not always the case. Some rapids become more technical and more difficult at lower water levels. The classification system also does not take into account the type of boat being paddled. Some rapids may present particular challenges for rafts, while other rapids may be more difficult for a kayak. As with any rating system, there is an element of subjectivity.

CLASS I: Moving water with a few riffles and small waves; few or no obstructions

CLASS II: Easy rapids with smaller waves; clear channels that are usually obvious without scouting; some maneuvering might be required

CLASS III: Rapids with high, irregular waves; narrow passages that often require precise maneuvering; scouting is often necessary

CLASS IV: Long, difficult rapids with constricted passages that often require complex maneuvering in turbulent water; the course may be difficult to determine and scouting is often necessary

CLASS V: The upper limit of what is possible in a commercial raft. Extremely difficult, long, and very violent rapids with highly congested routes, which should be scouted from shore; rescue conditions are difficult, and there is a significant hazard to life in the event of a mishap

CLASS VI: The difficulties of Class V carried to the extreme; nearly impossible and very dangerous; involves risk of life

reproduction or feeding. The Middle Fork of the Willamette is under consideration for increased work on floodplain restoration on publicly owned sites and where private landowners have an interest.

Moss clings to the rounded river rock along a Middle Fork gravel bar.

The Middle Fork below Dexter in places resembles the river of old, with an abundance of forested public land and a wild flair provided by the handful of rapids. Just downstream of the put-in at the Dexter hand launch, the first rapid appears. At higher flows this is a Class II rapid, with some standing waves that provide a short splash or two. Like any part of the river, it is the strainers that provide most of the danger here. You must take caution and paddle your canoe, kayak, or drift boat away from such immovable objects. Floods bring such strainers to the river and carry them away as well.

Dexter Launch to Springfield's Clearwater Boat Ramp, RM 203.5 to 191

⚠ This stretch contains three Class II rapids, as well as frequent snags. If you do not have experience in Class II rapids in your craft of choice, you have no business paddling this stretch of river.

STARTING POINT: Dexter hand launch. Take Jasper-Lowell Road from Springfield south along the Middle Fork to Dexter Dam. The launch is accessible via a driveway just before the dam.

ENDING POINT: In Springfield take 42nd Street south and take a left on Jasper Road. After about a half mile, turn right on Clearwater Lane to the Clearwater Park entrance.

DISTANCE: 12.5 miles

SKILL LEVEL: Class II whitewater in places, experienced paddlers only.

CONDITIONS AND EQUIPMENT: Use a whitewater boat and bring along necessary whitewater gear. There are three Class II rapids on this stretch.

AMENITIES: Pit toilets and free parking at both the put-in and take-out.

WHY THIS TRIP? The fast cold water and scenic riverside forest.

Put-in at the hand launch just before Dexter Dam. The current is strong at the launch site, and soon you encounter an island. To river left is Elijah Bristow State Park, with a lush riparian forest that hugs the west bank for 3 miles. Be very careful as you choose your route in this area. Strainers can be found nearly any time of the year in the channels around the islands, and occasional blockages of the channels can occur. How the river appeared the month or week before may not be how the river appears today.

After about 2 miles, there is a small rapid with nice standing waves. Again, watch for strainers on the outside bend. Further down you will encounter another rapid with a gravel shelf and occasional rocks. You should be able to identify a route through without too much difficulty. Throughout this stretch the river is scenic, though at times a bit too close to the sound of traffic on Jasper Road. You will be able to get a good look at Mount Pisgah from the east side. ▣

Elijah Bristow State Park

WHERE: Take Highway 58 southeast of Eugene approximately 7 miles. Signs point out the park entrance.

AMENITIES: Ample parking is available at the main parking area, as well as a map to help guide your visit. There is a day use fee.

This park is a wonderful large property along the Middle Fork of the Willamette, which has some 15 miles of hiking and equestrian trails. Standard day hiking equipment works well. The Oregon Parks and Recreation Department has been working on habitat restoration in riverside areas. In addition to planting projects, Elijah Bristow State Park is also a likely candidate for some level of bank modification to allow higher spring flows into the park, thereby replicating the river system's natural processes. Autumn is an excellent time to visit this park. ■

McKenzie River

The McKenzie River flows southwest from the Cascade Range into the Willamette mainstem. The average annual flow of the McKenzie River is about 450 cfs where it leaves Clear Lake and nearly 5800 cfs where it enters the Willamette River. Gaining much of its water from high mountain springs, the McKenzie's water runs pure and cold, especially as compared with other Willamette River tributaries. The McKenzie watershed is significant, at more than 1300 square miles, and the basin's topography is varied, ranging from mountain peaks of more than 10,000 feet to where the McKenzie reaches the Willamette at 375 feet. Although the McKenzie watershed has significant agricultural production, some 225,000 acres of the watershed is federally designated wilderness. The protection of the McKenzie River is critical, as it provides 200,000 people their drinking water.

As with the Middle Fork and Coast Fork, the McKenzie is manipulated by six hydropower dams that are owned by the U.S. Army Corps and the Eugene Water and Electric Board. These dams have greatly altered the river, as have the significant habitat modifications ranging from road building to water withdrawals. Estimates indicate that about 80 percent of the forest and related floodplain along the McKenzie have been altered by people. Only 8 percent of the length of the mainstem includes mature old-growth forest along both banks, and almost all of that is in the upper reach above Lost Creek.

In spite of these modifications, the McKenzie benefits from a federal law that provides some level of protection to river systems. The Wild and Scenic Rivers Act has listed the 12.7-mile stretch from Clear Lake to Scott Creek as a protected area, meaning that development is limited near the river. Sixteen miles of the McKenzie is also listed as an Oregon Scenic Waterway, with development curtailed within a quarter mile of the river in this area. Many people think that more miles of the McKenzie should have this designation, but some pristine sections are included in the current designations.

The McKenzie River benefits from an additional layer of protection known as the Three Basin Rule, which also applies to the North Santiam and Clackamas Rivers. The Three Basin Rule can be viewed as much more important than the Wild and Scenic Rivers Act in relation to overall river health. The rule requires additional consideration of any form of wastewater development in the pristine watersheds of these rivers, which provide drinking water to more than 17 percent of Oregonians. In 1995 the rule was improved with greater protection, prohibiting new or expanded wastewater discharges with-

A wayward sunflower on a gravel beach just downriver from the McKenzie confluence.

out specific findings from the Environmental Quality Commission and Oregon Department of Environmental Quality staff that any new facility will not degrade water quality.

The McKenzie is a special place for many people, whether they fly-fish, canoe, kayak, raft, or seek some of the many natural areas that surround the river. Like other rivers in the area, the McKenzie is home to spring chinook, summer steelhead, cutthroat trout, and bull trout, among others. The river has some good sized rapids as well, such as Martin's Rapid, a Class III rapid and a favorite of rafters and canoeists. Paddling the McKenzie can be an invigorating experience and can provide an abundance of information about the river.

Hendricks Bridge Wayside and Ben and Kay Dorris State Park

WHERE: Hendricks Bridge Wayside is 9 miles east of downtown Springfield on Highway 126, on the east end of the Hendricks Bridge across the McKenzie River. There is a boat ramp at the site that provides easy access for paddling or floating the river. Another land access point is Ben and Kay Dorris State Park, some 20 miles further east on Highway 126, 2 miles east of the town of Vida.

AMENITIES: Fee parking and restrooms at both sites.

These parks give a good sense of what much of the lower 30 miles of the McKenzie looks like. Both parks are open areas that provide easy access to the river. Hendricks Bridge Wayside is a 17-acre park that affords a view of the relatively tranquil McKenzie at this point. Ben and Kay Dorris State Park provides a good look at the McKenzie River just above Martin's Rapid, which is Class III. You can hear the rush of the whitewater from the boat ramp. Unless you have significant whitewater experience, you should not get on the river here. However, for those who want a better sense of the rapids, just to the right of the boat ramp there is a path that parallels the river and gets close to Martin's Rapid. ∎

In the 1930s there was a call to do something about the flooding of bottomland areas or floodplains. The U.S. Army Corps of Engineers responded, with their dam projects making a significant impact on the Willamette system. Private dams also exist on the McKenzie River, such as Leaburg Dam owned by the Eugene Water and Electric Board, whose three dams function only to generate electricity. The Corps, however, has had a long history with the Willamette River and has made a much more significant imprint on the river system overall than the few large private dams. Thirteen dams and reservoirs have been built and operated by the Corps in the Willamette Basin: Dorena and Cottage Grove Dams in the Coast Fork watershed; Hills Creek, Lookout Point, Fall Creek, and Dexter Dams in the Middle Fork drainage; Cougar and Blue River Dams in the McKenzie drainage; Fern Ridge Dam on the Long Tom River; Green Peter and Foster Dams in the South Santiam drainage; and Detroit and Big Cliff Dams on the North Santiam River. Multiple other dams were proposed by the Corps and local parties that were never built.

Based on local government support, a report was completed in 1937 by the Corps recommending a series of dams for flood control, irrigation, and other purposes such as municipal water use and recreation. There was some opposition to these projects, especially from those who understood how detrimental the dams would be to native fish. In the 1940s, however, the recommendations became law. The first two dams, Detroit and Big Cliff on the North Santiam River, were completed in 1953. Since that time the numerous dams have impeded the migration of native fish, a huge issue with animals that reproduce by returning to their natal spawning beds in creeks in the upper watersheds of the river system.

William L. Finley, the eminent conservationist, writer, and photographer, opposed the dam projects. Finley had served on the Oregon Fish and Game Commission and had acted as a commissioner, biologist, and game warden from 1911 to 1930. His photographic legacy was fantastic and helped to spur the creation of the National Refuge System. The William L. Finley National Wildlife Refuge was named in his honor after his death in 1953, the same year that the first two dams were completed. Finley and many residents along those rivers originally slated for the Corps dams opposed the projects on the basis of the impact to native fish, disruption to their communities, and the overall cost. He argued that alternatives were available and not enough research had been conducted. Finley also made the clear point that it was not the farmers in the valley who started the call for damming the rivers but rather boosters of the dams who developed the propaganda. Today one can question the costs versus the benefits of the dams.

The U.S. Army Corps of Engineers does indeed control flows in the Willamette system. The Corps has had to modify its operation of the dams to help accommodate for threatened species, with spring chinook leading the list of species of greatest concern. The manipulation of river flows, alteration of habitat, and blockage of migration routes makes a huge difference to these fish. The Corps has been required, at times reluctantly, to make changes to their projects and to develop ways to help reduce the impact of these dams.

The Coast Fork, Middle Fork, and McKenzie Rivers help to define the Willamette River, and some of the characteristics of rivers that can be found throughout the Willamette Basin. How these three tributaries are managed, manipulated, and appreciated by citizens greatly defines the overall health of the Willamette River.

On the
Willamette

1

The River in Eugene and Springfield, RM 187 to 175

We must be astute enough to see that preservation is far easier than correction, perceptive enough to realize that in the Willamette River we still have more to preserve than to correct, and bold enough to act accordingly.
—GOVERNOR TOM McCALL, 1967

EUGENE AND SPRINGFIELD are closely interwoven with the rivers in the area, with significant frontage along the Middle Fork, Willamette, and McKenzie Rivers. Because these waterways have good current and various levels of rapids, canoes, kayaks, and drift boats are widely used in the area. The McKenzie River drift boat has a curved bottom, a high arched bow, and a prominent stern, as well as its own lore and history of carrying fishermen across riffles and rapids. Most powerboats in the area are used at Fern Ridge Reservoir or the reservoir behind Dexter Dam. The connection of the area's residents to the rivers is also evident by the popularity of fly-fishing, with a surprising amount occurring on the Willamette just downstream of Eugene and Springfield.

While you can't objectively gauge how well a city or town has embraced its rivers, you certainly get the sense that a lot of people in Eugene and Springfield "get it." People who live in the area seem to understand the resources the river provides and how their actions can impact its health. They also seem to recognize what a wonderful natural feature they have flowing through their towns. However, the Eugene and Springfield areas also have a long tradition related to resource extraction, and this has certainly affected the Willamette and McKenzie Rivers. The waste of large paper mills and other facilities throughout the area have been dumped in these rivers.

The first white settlers in the area had some sense of how the southern end of the Willamette River system functioned. Rivers dominated this early settlement landscape, from the first cabin of Eugene Skinner situated on a hillside. *The Story of Eugene* (Moore et al. 1949) illustrated how early Euro-American settlements in the valley were connected to the river. The story opens with the

Previous page: Looking upstream, just downstream of Eugene. To the left, in 2007 the main river pushed through the island from which the photo is taken, providing a whole new course for the mainstem river. Now only a small channel heads to the right, where the main channel formerly was.

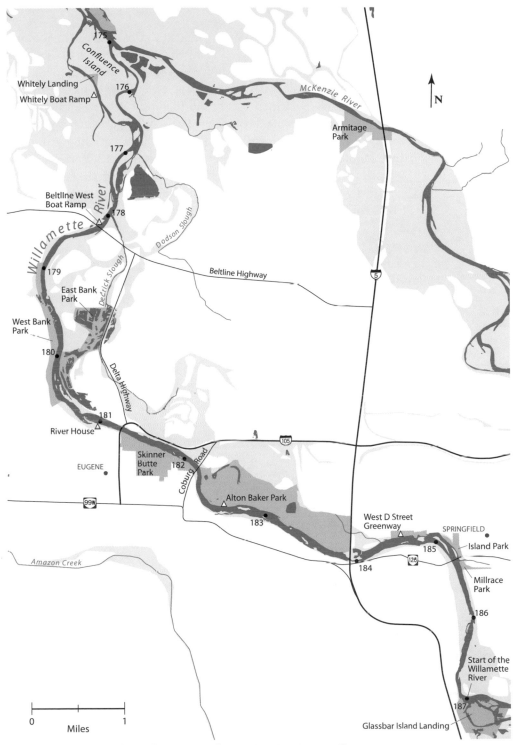

175

Confluence Island

Whitely Landing
Whitely Boat Ramp △

176

McKenzie River

Armitage Park

↑ N

177

Beltline West
Boat Ramp

178

Dodson Slough

Willamette River

179

East Bank
Park

Beltline Highway

I-5

Dedrick Slough

West Bank
Park

180

Delta Highway

181

River House △

Skinner
Butte
Park

182

Coburg Road

I-105

EUGENE

Alton Baker Park

West D Street
Greenway △

SPRINGFIELD

99W

183

184

185

126

Island Park

Amazon Creek

Millrace
Park

186

Start of the
Willamette
River

187

Glassbar Island Landing

0   Miles   1

Start of the Willamette to the McKenzie River confluence

image of Eugene Skinner standing on a rise next to the Willamette, looking out over the expanse of trees and water near what is today the city of Eugene. Two Native Americans come by as Skinner surveys the land for a cabin, trying to decide where to build. These Native people were said to have provided him some crucial advice, which was "build high up." When Skinner asked why, they responded "big waters come some day." While one may wonder at the accuracy of this story as described in the book, it is an account grounded in the functional natural history of the Willamette and its tributaries. Indeed, very big waters did—and still do on occasion—flow through the area.

The mainstem Willamette River is formed where the Coast Fork and the Middle Fork converge. The confluence of the two rivers just south of Eugene is a wonderful scene. To the southeast, behind the first few meters of the Willamette, stands Mount Pisgah. From the east the Middle Fork rushes northward, with relatively cold clear water from the Cascades, and from the south the less robust flow of the Coast Fork from the Coast Range meshes with the Middle Fork to form the Willamette River. At the confluence is a small public park owned by the Oregon Parks and Recreation Department. Just upstream is the Howard Buford Recreation Area, a major public park owned by Lane County. From the height of Mount Pisgah, at the end of a scenic trail from the parking lot, you can see the joining of the Coast Fork and Middle Fork. The view is well worth the hike up the long incline, as the summit provides a northwest-facing panorama of the Willamette's start and Eugene in the distance.

Aftermath of the flood of February 4, 1890, looking upriver from Skinner Butte. The covered bridge is near today's Coburg Road Bridge, and the land just upstream is the site of today's Alton Baker Park. Courtesy of the Lane County Historical Museum

The Willamette, like many other rivers, is a public river owned by the state. This idea has been somewhat controversial, as it relates directly to access to rivers and to the perception that a river's bed and banks can be privately owned. When a river is owned by the state (the agency responsible for administering it is the Department of State Lands), the river's bed and the banks up to the average high water mark are public domain. In addition, there is a floatability assessment in Oregon that gives anyone, anywhere, legal authority to travel any water body by boat if they can simply float it. This covers craft from drift boats, canoes, and even swimmers.

Public ownership of rivers has deep historical roots. When Oregon became a state in 1859, all the state's rivers that fell into the categories of "tidal" or "navigable" were considered owned by the state. In tidal areas, where the rise and fall of the tide occurs (such as the lower 26.5 miles of the Willamette), the tidal portion of the river up to the regular high water line is deemed public. The definition of a navigable river relates to its historic use for commerce, a rather broad-based term. If in 1859 the river was used or could have been used in its ordinary condition for commerce of trade and travel, the river is owned by the state. Commerce may have included floating logs from upriver to down, a boat transporting people from upriver to down, or a commercial ferry moving from one bank to the other. In some cases, use by Native Americans to harvest food and transport it has come into the commerce definition. Given the size of the Willamette River, this issue has not come to the fore, but navigability has been an issue for some of its tributaries.

When the state conducts a navigability study, it looks for the earliest evidence of commerce to help provide a basis for its conclusion. The state of Oregon, the Department of State Lands, and the Land Board have moved forward with greater purpose in conducting navigability studies on some of the state's rivers. A good recent example of this is the John Day River, which was ultimately deemed navigable after some significant conflict with a couple of landowners.

Some landowners have been angered by those seeking access to public waters, as they believe they own a river because it says so in their deed. In some cases, river travelers and fishermen have gone too far. In general, though, conflicts between landowners and river users have been few and far between. To avoid conflicts, you should not trespass on private property. The basic recommendation is to only walk into an upland area off a riverside if you know for sure it is public land. While navigable rivers should be used and appreciated, there is no excuse to trespass.

Throughout the cities of Eugene and Springfield, the Willamette River flows fast and relatively pure. Here the river provides a dose of whitewater and flat water interspersed through the length of the city, and the roar of Class II rapids can be experienced in the heart of Eugene. As a college town with

150,000 residents, Eugene has a good level of activity around the river on a regular basis. People flock to the Willamette throughout the summer months, when you can experience the splash and giggle of the inflatable raft full of kids or a tandem canoe getting doused with water in the series of rapids. For those who take the necessary level of precaution, the river in this area can be a lot of fun. Yet, as with any area with rapids, the bucolic setting can be deceptive.

Some people have had significant problems in this stretch because they have underestimated the Class II waves, assuming that they could take on the rapids with no experience and no lifejacket. Too often such assumptions, coupled with alcohol consumption, result in big problems. A good example of the dangerous conditions is the diversion dam at RM 184, just upriver of the Interstate 5 bridge. The low dam cuts across the river, leaving a gap at far right where a knowledgeable paddler can make it through. However, if you make contact with the dam and capsize on the downstream side, you could well be held under by a reversal wave that would rotate you against the dam. By stopping to scout the river beforehand, you'll see a flat horizon line ahead of you, with the water disappearing on the other side, an obvious sign of a dam, falls, or other structure. Numerous people have had difficulty with this and other diversion dams.

In the mainstem Willamette's first few miles, freeways cross the river, and highways and feeder roads edge along its banks. The start of the ubiquitous bank hardening begins where the river rushes under the Interstate 5 bridge. This hardening consists of rocks (or riprap) lining the riverside to control erosion. Riprap has been used extensively on many rivers around the United States. In places along the Willamette and its tributaries, riprap seems to be much overused. Healthy riverside areas are key to a healthy river, and riprap makes is difficult for native vegetation to thrive along the river. In cities like Eugene and Springfield, planting of riverbanks with trees and other riparian vegetation has increased as people seek to create wildlife habitat and to cool the water in riverside areas. Although the hardened banks detract from the beauty of the riverside area, the Willamette in this stretch is relatively clean, especially for an urban river, as measured by basic water quality parameters and riverside parks.

The city of Eugene has taken some positive steps for local waterways, especially tributaries of the Willamette. The restoration work along segments of Amazon Creek is a good example. Running along the west side of Eugene, this highly impacted creek had experienced the usual effects of urban development: loss of habitat and poor water quality. Since the late 1990s, the city and the U.S. Army Corps of Engineers have engaged in several restoration projects to improve habitat for fish and other wildlife and to diminish the input of pollution from runoff from nearby streets, parking lots, and related development. Eugene has also assembled a Stream Team that engages local residents in improving habitat along the city's interface with the Willamette and its tributaries (see Resources).

The increase in the Willamette Valley's population over the next several decades and beyond will present many challenges for the river's protection and restoration. As in other growing cities in the valley, Eugene's wastewater plant discharges into the mainstem Willamette. Typically the treated wastewater poses no threat to human health, but Eugene and Springfield are grappling with growth, leading to an ever-expanding need to treat more and more waste as housing developments and urban infill increase the population density of the area. A city of Eugene's size can affect the water quality not only through its sewer system but also from the stormwater runoff that carries substances from yards, parking lots, and streets. Eugene's Stream Team has done some very good work to disconnect downspouts and curb runoff into the Willamette. These actions can help prevent oils, phosphorous, and other forms of pollution from entering the river system.

Among the difficulties for Eugene and other urban areas is the need to keep river water cool enough to support native fish populations. Under the Clean Water Act, the Oregon Department of Environmental Quality must set standards for various water-quality indicators. The Willamette does not meet the standard for water temperature in the summer months, meaning the river is too warm. Water that is too warm can be detrimental to salmon and other aquatic species. As a result, the Department of Environmental Quality devel-

A 1912 image of the Willamette River at Springfield, taken from Willamette Heights. A bridge for wagons and autos is in the foreground, the Southern Pacific Railroad bridge in the center, and a streetcar bridge is in the background. Courtesy of the Lane County Historical Museum

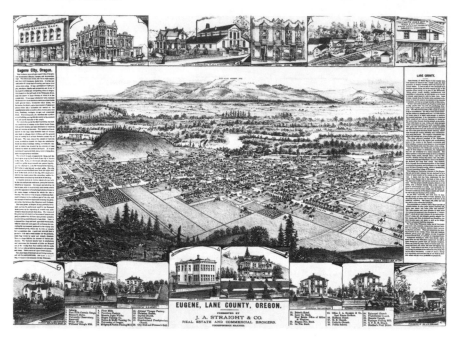

oped a plan to cool the river a few degrees, through actions such as planting shade trees along the river and its tributaries and enforcing a more stringent standard at water treatment plants. The theoretical total maximum daily load estimates how much heat the Willamette can safely accept without harming aquatic species (although this is based on an overall standard set by the state that not all agree with). Based on this, reductions in heat are required for industrial discharges, municipal wastewater treatment plants, and nonpoint sources (such as unshaded stretches of a creek or river). Because it is closer to the Willamette River's headwaters, water quality is generally better in the Eugene area, with cooler temperatures. Therefore, the Department of Environmental Quality developed more stringent requirements in this area.

In early 2007 the Metro Wastewater Management Commission, which is in charge of processing the waste of Eugene and Springfield, filed a lawsuit against the Oregon Department of Environmental Quality regarding these heightened standards. The commission believed that the standard was too stringent in the Eugene and Springfield areas as compared to other cities. Regardless of the difficulties involved, such as limiting urban and industrial development, cities like Eugene and Springfield must be part of the solution. In addition to keeping their effluent cool at the end of the discharge pipe, more riverside vegetation for shade needs to be planted. Ironically, the good work performed on Amazon Creek is exactly the kind of restoration work that is much needed throughout the rest of the Willamette Basin. Cities like Eugene may well pay for restoration work to offset their municipal discharges.

The Eugene and Springfield area provides an abundance of good opportunities to view the Willamette River and wildlife up close. For instance, on the pathway along the river at Whilamut Natural Area, in the spring you might witness a mother merganser leading her chicks behind the protection of overhanging branches of red osier dogwood. You can take in the river on the west side atop Skinner's Butte, which offers a panoramic view of Eugene, or at places such as Eastgate Woodlands in Alton Baker Park in Eugene and Millrace and Island Parks in Springfield. The Willamalane Parks system, the joint system of Lane County and Springfield parks, provides additional sites. A central opportunity to see the Willamette in Eugene is by riding or walking along the 12-mile Ruth Bascom Riverbank Trail that lines both sides of the Willamette River from the Willie Knickerbocker Bridge just west of Interstate 5 to the Owosso Bridge just south of the Beltline Highway. The trail affords an abundance of opportunities to see the rippling waters of the Willamette. In addition, Eugene's Outdoor Program has a River House at Maurie Jacobs Park, which offers instruction for canoes and kayaks and continuing trips and education courses related to river travel.

Eastgate Woodlands–Alton Baker Park, Island Park, Millrace Park, RM 185.5 to 183

WHERE: Along the Willamette in Eugene and Springfield
AMENITIES: Parking areas, restrooms, picnic areas, playgrounds, and a boat landing at Alton Baker and Island Parks; no facilities at Millrace Park.

The 40-acre Eastgate Woodlands runs along the Willamette at the eastern end of the Whilamut Natural Area of Alton Baker Park, a greenway park connecting Eugene and Springfield. The woodlands has two main trails: the Riverside Trail winds along the river's edge through red alder, willow, and black cottonwood trees, and the Woodlands Trail travels along a canoe canal through a shady bigleaf maple forest. A great blue heron colony nests within the woodlands from February through June each year. The 400-acre Alton Baker Park has an amphitheater, picnic areas, and some good fly-fishing opportunities.

Millrace Park is a 0.7-acre site across the street from the old train depot that serves as the Springfield Chamber of Commerce. Despite its small size, deer, river otter, great blue heron, green heron, spotted sandpiper, northern flicker, cedar waxwing, bushtit, ruby-crowned kinglet, and black-capped chickadees can be seen in the park. The 14-acre Island Park is adjacent to a set of mid-channel islands at RM 185. This park offers a boat landing, fishing, trails, and river viewpoints in the heart of downtown Springfield. ∎

Rivers maintain a good part of their health in relation to what happens on the land surrounding them. Because of this, protecting and restoring connections between wetlands and local tributaries to the Willamette are critical, especially in efforts to provide native species a place to flourish in an urban area. In 1992 the city of Eugene created the West Eugene Wetlands Program to address many issues, such as the need for a mitigation bank to protect and restore wetland habitat as more development occurs in the city. The West Eugene Wetlands is perhaps a bit different than most wetland areas in cities, given its relatively large size. The program represents a strong partnership that restores wetland and stream habitat and provides river access, wildlife viewing, and educational opportunities. It also operates a native seed collection program to provide seed of locally native wetland, riparian, and upland species to use in restoration projects. Amazon Creek benefits from the program, as it is adjacent to some portions of the wetland. Further, a land trust partnership has been developed between the Bureau of Land Management and the city to acquire additional land from willing sellers.

Mitigation is the process of preserving, enhancing, restoring, or creating habitat to compensate for consequences of land development. The mitigation bank of the West Eugene Wetlands Program allows those who are developing land and compromising other wetland habitat to purchase credits for habitat at West Eugene Wetlands. Like other mitigation banks, the credits have been calculated by government sources and are related to the breadth and quality of habitat established at the wetlands. The aim of the program is not to let developers off the hook through mitigation, but to provide better wetland habitat in a large area that offers greater benefit for area wildlife (as opposed to creating a small marginal project adjacent to the developed area, which provides less benefit).

Like Eugene, Springfield is working to restore the ecological function of the city's river habitat as well. In close cooperation with the Middle Fork Watershed Council and other partners, the Springfield Utility Board has actively addressed habitat restoration and ways to cool river water during the summer months, in an effort to improve habitat for native fish.

Both Springfield and Eugene have partnered with the U.S. Army Corps of Engineers to address floodplain restoration on the Coast Fork and Middle Fork. Along with other municipalities, the cities have participated in a study of how controlled flooding in floodplain area can create valuable habitat and help to decrease the severity of flood events. In many cases, floodplains have been separated from their rivers. These areas are no longer able to absorb water, thereby decreasing the overall flood height. Though no projects have been implemented, the fact that Eugene, Springfield, and the Corps are looking at such options is a good sign, for both the protection of property and the improvement of the natural ecological function along the rivers.

As you will readily see if you travel the Willamette, Middle Fork, Coast Fork, and McKenzie Rivers, numerous gravel bars exist along the way, with some of the best deposits located right along the river. Over the millennia, the rushing waters of rivers like the Willamette and McKenzie have carried rocks from high in the mountains, rounded them over the miles and years, and deposited them across the broad historic floodplain, distributing layer after layer of gravel in the river, along the river, and outward from the river.

Gravel is necessary to make concrete, which is essential for the construction of roads and buildings. Many gravel mining companies are located along the Willamette in Eugene, particularly between RM 178 and 175. From the air, as one looks northward from near the confluence of the Willamette and the McKenzie, the river almost seems lost amid a wide expanse of brown, blue, bluish green, and dark water-filled gravel pits. In fact, when flying over the majority of the Willamette River mainstem, one is rarely out of the sight of gravel pits right next to the river. The evidence of these operations is plain to see. Due to the construction of dikes at the gravel extraction sites, the riverbanks tend to be taller and denuded. Trucks and other machinery can be heard at a distance, which can make such areas less than appealing to wildlife.

It is common for gravel mining companies to purchase land along the Willamette and other rivers and then seek a zoning change from agriculture to mining at the county level. For decades, the permitting structure provided by counties and the state of Oregon for such gravel extraction operations was not very robust or transparent, making it difficult for those concerned about the river to track the development of such operations. Since 2002 both state and federal natural resource agencies have given additional thought to how these operations, with deep pits so close to many rivers, should be evaluated.

When the local community has weighed in and provided enough pressure for the county to carefully evaluate the zoning change, interesting issues have come to the forefront. For instance, local residents often react to the idea of hundreds more trips by large trucks along local roads, in a ceaseless march back and forth from quarry to processing plant. In the Eugene area, this traffic has been a heated issue as operations have been proposed in the last few years. In addition, when large gravel pits are dug, the depth of a pit may affect the surrounding groundwater, thus lowering the water table for the wells of nearby landowners and potentially affecting the inflow of fresh groundwater into the Willamette River. Areas where groundwater flows into a river in significant volumes are known as hyporeic zones. This type of groundwater flow can be warmer in the spring, aiding the hatch of river insects, and cooler in the summer, providing critical refuge for fish in the hot summer months. In addition to these effects on wildlife habitat, gravel extraction operations clearly affect the aesthetic quality of the river. Although some have proposed that the future reclamation of gravel pits might provide benefits for wildlife—

if reconnected to the river these ponds could provide habitat for native fish and other species such as western pond turtle—the fact is that a gravel extraction operation is unsightly.

When you encounter these gravel extraction pits along the river—whatever value they seem to provide to some people in the name of progress—their presence in the Willamette floodplain is but one more sign of what we have taken from the river. At the very least, proposed gravel extractions operations and expansions to existing ones deserve careful consideration related to clean water, impacts on wildlife, and the lifestyle in the surrounding community. Because the population of the Willamette Valley will continue to grow, the need for concrete will continue to grow as well. Therefore, it is important for federal, state, and county management agencies to comprehensively understand the effects of gravel mining along the Willamette now and in the years to come.

2
Eugene to Harrisburg
RM 175 to 161

The path of the paddle can be a means of
getting things back into their original perspective.
—Bill Mason

THE STRETCH OF RIVER between Eugene and Harrisburg is well suited for paddling. The legendary Bill Mason, Canadian canoeist and naturalist, really knew how to capture the essence of rivers and traveling on them. His philosophy is that floating along in a canoe, kayak, drift boat, or some other low-impact means is perhaps the best way to experience and understand any waterway. Intimacy with the flow of the current or the gentle shift of light on the bottom of a deep pool is readily captured at the pace and perspective provided by these craft. When you dip your paddle into the water and move into the flowing river, new perspectives can open up in front of you. On the Willamette the dynamic interplay between the shifting current, the ever-changing channels, and the rotation of the seasons means that no two paddle trips are the same.

This chapter explores one of the liveliest stretches of the Willamette River and one that most closely resembles the river of a century ago. From the ever-changing shape of the river to the significant stands of riparian forests, river travelers will likely not be disappointed along this stretch, from where the McKenzie joins the Willamette in Eugene to Harrisburg. Along this route, the story of the Willamette's ecological issues also unfolds, with one of the most prominent—the loss of historic channels and floodplain habitat—being close at hand.

The Willamette River is in fairly good shape as it flows from Eugene and Springfield, although the river has encountered the effects of industry, storm-water runoff, and wastewater that it will meet again and again as it flows northward. For the next 14 miles to Harrisburg, however, the riffles, islands, and fast-moving current provide a semblance of the river of old. At Whiteley Landing, a small put-in on a scenic backchannel just west of the McKenzie confluence, the river flow is shallow and clear. Even in the midst of the Eugene suburbs, you can sometimes see the lithe motion of a mother deer leading her fawn across the shallow ripples of the backchannel, seeking refuge on an

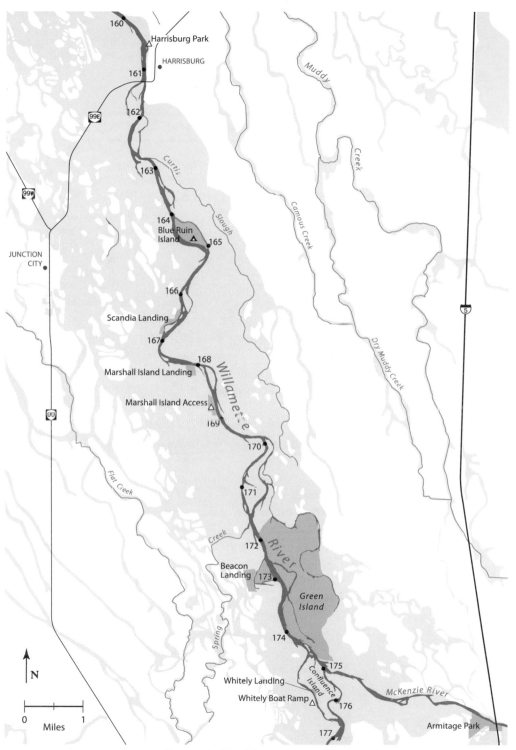

160

Harrisburg Park △

HARRISBURG •

161

99E

162

Curtis

163

164
Blue Ruin
Island △ 165

Slough

166

Scandia Landing

167•

168

Marshall Island Landing

Willamette

Marshall Island Access △

169

170

171

172

Beacon
Landing △ 173•

River

Green
Island

174

175

Whitely Landing

Confluence Island

Whitely Boat Ramp △

176

McKenzie River

Armitage Park

177

JUNCTION
CITY •

99W

00

Flat Creek

Creek

Spring

Muddy

Creek

Camous Creek

Dry Muddy Creek

5

N

0   1
Miles

Eugene to Harrisburg

early summer morning. This backchannel and the adjoining island have what was once common along the river, multiple channels and low gravel islands full of trees.

Whitely Landing to Marshall Island, RM 175 to 169

STARTING POINT: From the Beltline Highway in Eugene, take River Road north to River Loop Road. Follow River Loop 1 east to Wilkes Drive. Take a right on Wilkes and follow it to the boat ramp at Whitely Landing.

ENDING POINT: From the Beltline Highway, take River Road north approximately 8 miles to Hayes Lane. Take a right on Hayes and follow it to the Marshall Island Access park entrance.

DISTANCE: 6 miles

SKILL LEVEL: Experience on meandering channels with strong current.

CONDITIONS AND EQUIPMENT: This put-in is along a shallow backchannel that during low-water months is best suited to canoes and kayaks. Watch for gravel and occasional strainers. These channels can change with the seasons; new river features, such as logs, can appear one day and be gone the next. Remain watchful as you travel by paddle craft.

AMENITIES: Fee parking at Whiley Landing, free at Marshall Island. Pit toilets at both locations.

WHY THIS TRIP? You can view one of the Willamette's most dynamic areas and get a sense of what backchannel habitat used to look like along the river.

Whiteley Landing is on the backchannel side of Confluence Island, where the McKenzie joins the Willamette. From Whitely Landing, the current at most flows is pretty mild. This backchannel is relatively shallow, between approximately 1 to 10 feet in depth. Along its 1.5-mile length, where it meets the mainstem, the water flows relatively fast over large gravel, interspersed with a few riffles and a small pool section toward the end. The small channel connects with a lush riparian zone in some areas, with mature cottonwoods along the lower section, as well as small backwaters and related eddies. It is not uncommon to see deer sipping at the edge or cedar waxwings in the spring and summer darting from tree to tree in the riparian fringe.

As the main channel emerges, the left bank with a mixture of cottonwood and willow is owned by the Oregon Parks and Recreation Department (Rogers Bend Landing, RM 174.5). Green Island, owned by the McKenzie River Trust, is on river right. This island is being restored to provide habitat for fish and wildlife, with parts of its riverside lands and upland areas being replanted and side-channel reconstruction in the works.

At RM 173.5, there is a slight depression as the wall of riprap recedes and a change in elevation that appears to have been a backchannel or an area that was routinely flooded. This upward portion of the river's left channel would have been a natural backwater suitable for a wide range of wildlife. Yet at high spring flows such flooding would have threatened the use of the adjacent agricultural lowland. Therefore, the mainstem river was channelized and flow was restricted from side channel areas.

The river divides near Beacon Landing. The flow to the left is the new channel.

The Willamette divides around an island at RM 171.5. The primary channel has changed from east to west and back over the years. Note the nice backwater area in the middle of the island.

Just past RM 173, Beacon Landing appears on the left. Today, the river flows due north, not to the west, as it had in prior years. As you pass the Beacon Landing area, the islands to the left and right were one large island, which had a small channel that received flow only during high-water periods. In the winter of 2005–2006 the channel grew larger, and then in early 2007 it pushed completely through. This change in course reflects the power that the river can still have in these dynamic areas when winter and spring flows are high. An immense volume of soil, rock, and trees was moved downriver.

Further downstream, power lines come into view. On river left among the cottonwoods and occasional snags, bald eagles can be seen frequently. These great birds are typically perched in the upper branches, but they can also be found closer to the ground as well on woody debris.

Just ahead at RM 171.5, the channel divides around an island. The main channel goes to the left, with a lesser channel of approximately 20 percent of the total river flow heading to the right. The right channel can be navigated, but use caution and watch for snags given that the channel can change from season to season. This area is very scenic. Looking toward river right you can view the broad river bottomlands with cottonwoods and willows and to the east the Coburg Hills in the distance. The left channel has significant flow and velocity. In 2004 high flows blew out an immense tangle of logs, root wads, and other woody debris that had clogged the channel opening, scattering the

material throughout the next mile of the channel. Consequently, river travelers must remain focused on what is downstream. In 2007 a portion of the left channel's lower stretch was clogged with woody debris, which even for the experienced canoeist required careful maneuvering.

As you pass the island, the river gently turns to the right, straightens out, and turns to the right again. The Oregon Parks and Recreation Department's Marshall Island Access is on river left just past RM 169. The simple boat ramp or the small gravel area to the right of it make the best take-out. ▣

The boat ramp and adjoining gravel beach at Marshall Island Access make a good put-in and rest stop along the Upper Willamette.

The mainstem Willamette River was channelized by armoring its banks with riprap, large rocks that prevent the opening of side channels. In addition, large wood structures are also erected to work in tandem with riprap to keep the bank in place. This process vastly decreases the natural meandering of the river during the annual spring floods, thereby curbing the productivity of the river itself. It also greatly restricts the health of the riparian area, with large rocks placed where willows and other plants would otherwise be found. Placement of riprap along the Willamette is one of the key actions that has disconnected the river from its historic floodplain areas.

From Eugene to Harrisburg, the Willamette is relatively shallow, with many areas of fast-flowing water and numerous riffles. This dynamic stretch also contains many islands and side channels that can change significantly from year to year. In the past, far more side channels connected to the floodplain along the Willamette, especially from Eugene to Corvallis. Side channels spur off from the main river channel, running roughly parallel, then reconnecting to the main river, sometimes miles downstream. When unconstrained by riprap, such channels also help to inundate floodplain areas during high flows. Both floodplain side channels and side channels that are constrained within the overall river channel exist in the Willamette system.

Historically, the natural force of the Willamette pushed great volumes of water across the landscape in the late winter and spring. Consequently, low-lying areas adjacent to the river became secondary channels that nourished floodplain wetlands. Flooding—a natural river process—was common. In the 1870s when the southern Willamette Valley was experiencing rapid settlement, the effort to separate side channels from the mainstem in some cases affected miles of backchannel area that would have flowed back into the main channel miles downriver. The technology was simple, as it is today. Pilings were driven into the river bottom at the channel openings, and rock was placed behind them. As these walls were erected, the water was slowly diverted back toward the main channel. Wing dams are also constructed of similar material; these walls are not at channel openings, but instead are placed wherever people wanted to direct water away from the shoreline and

Pilings in combination with riprap rock line many stretches of the Willamette, helping to restrict the flow of the river to the main channel.

back into the main flow. With the flow unable to make its way into the channel, unfortunately many of these channels dried up.

Surveys conducted by the U.S. General Land Office in 1875 identified the wide array of channels that formed the Willamette River. You can imagine the early surveyors, trudging through lowland brush, within earshot of the river, stumbling through a green fog of willows to plant their feet into the fine silt along the bank. They may have looked up to see a fast-flowing rivulet channel, staring across a shallow, muddy riverbed to another barrier of willows beyond. Somewhere to the side a heron may have croaked as it lifted off a cottonwood snag, startled by the surveyor's efforts to move forward. All around them, they would have experienced the sound of water, bursting and bubbling against rock and trees, with no clear end in sight. John Creath Bramwell, who was born in Halsey, Oregon, in 1857, noted, "Before drainage ditches, the whole valley was like a swamp. The streams, many of them after leaving the foothills, had no definite channels but spread out over the floor of the valley, wandering here and there over all the land." This vision was likely commonplace in many areas of the Willamette River system.

These channels at times opened to wider floodplains along the floor of the valley, extending water far inland and inundating the vast expanse of lowland forest and wetland areas. Natural floodplain function of this type was critical to a host of species, ranging from fish, mammals, and birds to turtles and amphibians. Such areas also provided a linkage for water to the river, whether via their hydrologic connection to side channels or the passage of groundwater and surface water to the mainstem Willamette.

In common usage, the word *revetment* refers to the assemblage of structures to block backwater areas from the main flow of the river, and 96 miles of revetments have been constructed along the Willamette. Just below Marshall Island Landing is an example of such a revetment. The bank has large wooden pilings driven into it at RM 168, about 10 feet from the river, with the bank itself holding a mass of riprap leading down into the water. In fact, remnants of many piling structures can be seen today along the mainstem Willamette. In some cases, these continue to restrict the flow behind existing islands. In other cases, though, they provide a hazard to canoes and kayaks as well. Any fixed structures in the water, whether pilings or snags extending from the river's edge, are a potential hazard to paddlers. Newer bank hardening efforts can also be seen. Along the stretch of the Willamette River just above Harrisburg, much of the bank has a large wooden fencelike structure and riprap, and, most of Harrisburg's waterfront is riprap.

In the historical context, we can probably understand why people made such modifications to the river. After all, early residents experienced a vast river that frequently flooded their crops and homes. While the long-term ecological consequences were not of primary concern, the overall change in

Riprap is common along the Upper Willamette, keeping the bank in place and flows out of historic floodplain areas.

channel complexity from these human modifications has been dramatic. According to analyses conducted by the Pacific Northwest Ecosystem Research Consortium, what was more than 41,000 acres of braided channels in 1850 had decreased to fewer than 23,000 acres by 1995. In 1850 this stretch between Eugene and Harrisburg contained nearly 25,000 acres of river channel and islands, whereas today it stands at only 8000 acres. Numerous changes of this sort have corresponding impacts on wildlife. The numbers of fish, mammals, and birds have declined, which can be clearly linked to the alteration of the river's natural function.

The braided network of channels that had moved water northward from Eugene was slowly transformed into the single primary channel that is seen along most of the Willamette today. This gradual shift from a complex assemblage of channels that grew, changed, and meandered over the seasons, supporting a wide array of native species, has triggered major changes in the overall health of the Willamette River. Today, native spring chinook and winter steelhead are threatened species, while introduced species such as smallmouth bass thrive in this altered system.

One can imagine the rise and fall of the historic channels with the seasons, with big brown swirling masses of springtime water shooting down the Coast Fork, Middle Fork, and McKenzie Rivers, surging onto the relatively open expanse of the valley, spreading outward from the main channel through thick riparian vegetation, and releasing northward along the contours of the valley floor. Such force moved and shaped the river channels, morphing the endless expanse of rock, building gravel bars, and honing the arc of the channels. The surge of water could cut through the layers of gravel, silt, and clay, creating new braids among the cottonwoods, willows, and ash. These areas would then expand and contract, all the while providing additional habitat for fish and essential habitat for some birds and mammals. All along the Willamette River a dynamic connection between water and land nurtured native plants and wildlife. The loss of thousands of acres of lowland habitat, the myriad backwaters and sloughs teeming with rippling currents and sedentary backwaters, eliminated an enormous amount of river habitat for native plants and wildlife.

The imagery of the historic river is important to capture, for there is some hope that we may one day regain some of the river's function. In fact, this stretch of the Willamette between the McKenzie confluence and Harrisburg reflects this possibility. Here there are larger tracts of lowland, some of it public land, that have been blocked off from seasonal flooding. Some of these lost backwaters and channels could indeed be reconnected to the river, benefiting both people and wildlife.

Remnants of channels that once raced parallel to the main current, old oxbows, and backwater areas can still be recognized along the river. From the air, their contours can be seen in areas now farmed for grass seed as well as in unplowed vegetated areas. During the flood of 1996, the Willamette River temporarily reclaimed many of these areas. The massive flooding made it clear that the river would find its floodplain, developed or otherwise, and that the lack of natural floodplain habitat had dire consequences for towns and cities along its banks. While the river is controlled extensively by the U.S. Army Corps dams, it still has the ability to change.

One of the chief aims of conservation groups is to develop restoration projects along the Willamette River that can regain some of the prior channel complexity and functioning floodplains. A fantastic example of how this might occur is at Green Island. This wonderful riverside property stretches along the lower 2 miles of the McKenzie River and then down the east bank of the Willamette for another 3 miles. In 2003 the McKenzie River Trust, with grants and contributions provided by state, federal, and private sources, purchased the land from a private landowner. Green Island consists of nearly 1000 acres of floodplain lands, and it once hosted one of the McKenzie's main channels. Today pools from this remnant channel can be seen here and there among the low-lying areas of the property. Other channels may have made their way through the property seasonally. As with many other portions of the Willamette River system, this former floodplain was separated from both rivers and utilized for agriculture. The McKenzie River Trust and its partners are working with state and federal natural resource agencies to examine the potential of increasing channel complexity by opening up portions of the Green Island property to the McKenzie and Willamette Rivers. This would create new river habitat.

While this is but one property along the Willamette, it provides a ready example of the conservation and restoration potential along the river. The land was purchased from a private landowner using public and other funding, and today the natural condition of the area is being improved. Along the 187 miles of the mainstem Willamette are other opportunities to conduct similar work, with the river and the public deriving significant benefits. Willamette Riverkeeper, partnering with the Oregon Parks and Recreation Department and the Oregon Watershed Enhancement Board, is working on similar projects at greenway parks.

A mature black cottonwood in a classic pose at the river's edge.

Black Cottonwood

The black cottonwood (*Populus balsamifera* ssp. *trichocarpa*) is a common deciduous tree along the Willamette throughout its extent. Although Oregon ash (*Fraxinus latifolia*), Pacific willow (*Salix lucida* ssp. *lasiandra*), and other trees are present in the lowland forests, the black cottonwood forms much of the riparian and lowland forest along the river, occurring from a riparian strip a couple of trees deep to sizeable stands in greenway parks and some private lands along the river.

Most anywhere along the Willamette River or lower sections of the Middle Fork, Coast Fork, or McKenzie, you can paddle up to a gravel bar and find a place to sit next to a long horizontal downed tree resting on the rounded gray stones sculpted by the current. It is another black cottonwood tree that has succumbed to the erosive force of the river. Cottonwoods are slowly recruited from the riverbank, falling into the current and perhaps ending up across the river a couple of miles downstream, in some cases the roots still clinging to the rich dark soil and large rocks that was the former riverbank. When the water rises again for a sustained period, these snags could be swept downstream again, perhaps with their gnarled mix of roots catching a gravel bar in the middle of the river.

A quick scan along an outside bend of riverbank reveals cottonwoods in varying states, trying to avoid the day they might come crashing down into the river as the result of a high spring flow or perhaps only a gentle wind. The trees' root wads are exposed to the great force of the river, which over time strips away the shoreline until there is more air than soil around the roots. In such cases, a slight push of wind and the forces of gravity can send the giant tree crashing into the swirling current. When a whole cottonwood recently peeled from its grip on the bank comes floating past at 10 mph, it should make a canoeist stop and think about getting on the river. Although this regenerative process of erosion and addition of wood to a river system is amazing, this debris can provide significant hazards for any river user.

Black cottonwood bark has a deeply grooved texture.

Black cottonwoods thrive on the disturbance of a changing riverscape, which enables new trees to become established along sandy, gravel shores. Researchers Bruce Dykaar and P. J. Wigington have looked into cottonwood regeneration along the Willamette. They compared aerial photos going back to 1936, well before the dams were constructed, to images from recent times and noted a significant decrease in the prevalence of cottonwoods along the river. Some of the change correlates with the decrease in flows after the dams were constructed and channelization of the mainstem, which reduced the number of gravel bars and islands that promote cottonwood establishment. The authors concluded that building dams and riprap revetments, mining gravel, clearing forest, and expanding agriculture had substantially altered the river environment needed for healthy cottonwood stands. Considering the importance of cottonwoods in the Willamette system, this species merits further restoration efforts as well as additional research on how flow management by the U.S. Army Corps affects this and other riparian and lowland forest species.

Marshall Island Landing to Harrisburg, RM 168 to 161

STARTING POINT: From the Beltline Highway in Eugene, take River Road north approximately 8 miles to Hayes Lane. Take a right on Hayes and follow it to the Marshall Island Landing park entrance.

ENDING POINT: From Eugene, take Highway 99 north to Harrisburg. Harrisburg Park is located on South 1st Street.

DISTANCE: 7 miles

SKILL LEVEL: Experience on meandering channels with strong current in places.

CONDITIONS AND EQUIPMENT: Recreational and sea kayaks, canoes, and drift boats work well along this stretch. Conditions change and woody debris is common.

AMENITIES: There is free parking at both points, a pit toilet at Marshall Island Landing, and restrooms at Harrisburg Park.

WHY THIS TRIP? This is another wonderful open stretch of the Willamette with multiple side channels that shift from year to year and season to season.

A classic gravel bar island, Blue Ruin has a nice backwater for camping.

At the put-in at Marshall Island, an Oregon Parks and Recreation boat ramp, the flow is quick and soon leads past an island owned by the Department of State Lands. In the spring you may witness willow flycatchers (*Empidonax*

traillii) perched on Pacific willow in the riparian fringe. According to *The Birds of Oregon* (Marshall et al. 2003), these birds have become less numerous in recent years but can still be found along the Willamette.

At RM 165 is Blue Ruin Island, a large gravel island interspersed with stands of Pacific willow and replete with wildflowers in the spring. It has a small backchannel, with the main current moving over gravel bars adjacent to the island, arching leftward, then on past the island. Around the far end is a Willamette Water Trail campsite and an area of standing water that works for a take-out. The side channel flows past, ending in shallow riffles in spring, but with very little water during the summer months. A visitor might see bald eagles flying overhead and osprey defending their catch, swooping down toward the eagles. While some of these gravel islands look slightly bleak, they often hold wonderful scenery and hidden wonders and are worth a stop and some exploration. Bald eagles often can be seen downstream of Blue Ruin Island as well. They have nearby nests and perch high in the riverside cottonwoods.

Between RM 163 and 164, on river left there is a large gravel area, interspersed with Pacific willow and black cottonwoods. This area is very low, with contours indicating that water spreads throughout during periods of high flow. In essence, this is a floodplain and likely provides some key function for wildlife.

Around a bend a railroad bridge crosses the river. Swirlies, areas where the river shifts the current back and forth with eddy lines, are found just under this bridge. These currents are harmless but can move a craft back and forth a bit.

Snags and woody debris are common in side channels, such as this channel that opens on the upper end of Blue Ruin Island.

Looking to the right, just past the railroad crossing, a small side channel flows off from the main current and delineates a large gravel island. In 2006 the city of Harrisburg made plans to acquire the land to the east behind the island for a new natural area.

The final mile of this stretch leads into Harrisburg, where the Highway 99 bridge crosses the river. The right bank in the city is composed of large riprap.

The railroad bridge above Harrisburg

A cottonwood snag on the Harrisburg gravel bar

As the gazebo comes into sight, the boat ramp is just downstream. In 2005–2006 high flows brought in a massive gravel bar at the base of the boat ramp, which made access for boat launching difficult. However, the new gravel bar stretches downriver for approximately 300 yards, providing ready access to the river. In the past, the only access was via the existing boat ramp or occasional openings in the riparian vegetation downstream. The city received a grant from the Oregon State Marine Board in 2005 to construct a new boat ramp. Because of the popularity of the gravel bar, the city plans to leave much of it in place for public access and excavate just enough to develop the new ramp. ▣

Osprey

From March through October, osprey (*Pandion haliaetus*), sometimes referred to as fish hawks, can be seen all along the Willamette River, but especially in this stretch. This magnificent bird, with a whitish head and underparts and a dark brown eye stripe and wings, frequently hovers over the river's surface, scanning for fish. When it sees one, the osprey dives at amazing speeds with feet and talons extended, hitting the water feet-first with a splash, sometimes becoming almost completely immersed in the water. Barbed pads on the soles of the bird's feet help it grip slippery fish. If the hunt is successful, the osprey lifts with labored wingbeats, carrying the fish to a nearby perch to feast on. The bird sometimes adjusts its catch numerous times, always ending up with head pointing forward for the best aerodynamics. On the Willamette, the fish of choice for the osprey is the native large-scale sucker (*Catostomus macrocheilus*).

An osprey soars against the riverside cottonwoods.

Frequently mistaken for bald eagles, osprey are approximately two-thirds the size of eagles with a different pattern of dark and light coloration. Their heads are distinct, with the usually light-colored head halved by a dark eye stripe (whereas mature bald eagles have the characteristic solid-white head). Osprey make their nests at the top of exposed snags or, more frequently, on platforms placed on the top of power poles. Their high-pitched cries can be heard, especially when they are near other osprey. North American osprey make an amazing annual journey, traveling far south into Mexico or beyond. During the winter months, Willamette River osprey go as far south as the Yucatan Peninsula.

Today osprey are common along the Willamette and many of its tributaries, but this was not always the case. The birds were frequently shot by fishermen, as starkly illustrated in *The Birds of Oregon* by Gabrielson and Jewett (1940), "The Osprey, or Fish Hawk, formerly common along the Columbia and Willamette Rivers, in the Klamath basin, and about the large Cascade lakes, must now be considered one of the rarer Oregon hawks." The authors noted protections in many other states at the time, "but in Oregon, like all other hawks, they are killed at every opportunity, both by farmer boys and those sportsmen who begrudge them the few fish they consume."

Later studies by Chuck Henny of the U.S. Geological Survey in Corvallis and others indicated that conversion of habitat and the significant use of DDT on agricultural crops hurt osprey populations as well. DDT is a bioaccumulative toxin—meaning that the chemical is not degraded as it moves up the food chain—that causes the thinning of birds' eggs. This egg thinning brought about a drastic decline in successful reproduction and subsequent decreases in osprey populations along the Willamette. In 1972 the use of DDT was banned nationwide, which likely has contributed to a rebound in the number of osprey and other birds of prey. In addition, the placement of platforms on power poles has also helped the state's osprey population increase. Significant cooperation by power companies has enabled numerous platforms to be installed along the Willamette and many other areas in Oregon.

Osprey remain a species of significant interest. Because predators bear the brunt of bioaccumulation, osprey could once again be affected by contaminants in the Willamette and the fish that the birds consume. As a result, it is necessary to monitor birds of prey to ensure that they are not being negatively affected by newly found contaminants in the Willamette system. A good example of the potential ongoing risk to these birds is polybrominated diphenyl ethers (PBDEs). These compounds have flame-retardant properties and are integrated into the manufacture of electronics, curtains, and other household items. As these products are disposed of, the PBDEs are transferred into the environment. Like PCBs and DDT, PBDEs are bioaccumulative compounds. PBDEs are also showing up in increasing levels in people and are listed as a potential carcinogen by the U.S. Environmental Protection Agency, generating increasing concern about the prevalence of PBDEs in the environment. In 2004 the use of two of the worst PBDE variants was voluntarily discontinued in the United States, but the use of deca-BDE continues. In 2007 the Washington State Legislature took a bold step by enacting a ban on the sale of most products containing PBDEs. This was possible because safer alternatives to PBDE exist and there is a growing understanding of the potential health and ecological risks. Hopefully the state of Oregon and the federal government will follow suit and enact similar legislation.

The final link on this stretch of the Willamette River is the city of Harrisburg, with just over 2000 residents. The town was founded in 1852 and was originally called Prairie City. In 1866 it was officially incorporated as Harrisburg. Early on, steamboats made their way to Harrisburg to pick up agricultural trade goods and transport people, although the river's fast and shallow flow made steamer travel challenging. Today, the surrounding area is largely agricultural, with some industry related to gravel extraction. Many of the city's residents commute to Corvallis, Eugene, and Junction City for work.

Harrisburg maintains a nice park along the river, offering a gazebo, restrooms, and running water. The boat ramp is linked with a large gravel bar, and improvements to the ramp's condition are planned. Since 2000 the city has reoriented toward the river, understanding opportunities for tourism and the economic benefits that local stores and restaurants can receive. Harrisburg has welcomed those who drift boat, fish for cutthroat trout, and paddle from Eugene and beyond. This park is also a vital link on the Willamette Water Trail, an effort well supported by Harrisburg. In the years ahead it is likely the town's connection to the Willamette will strengthen further.

3

Harrisburg to Peoria
RM 161 to 141

Since this nation is already loaded with debt,
why shove our nation more into the red by doubling
the debt for this flood control project?
—WILLIAM FINLEY

HARRISBURG PROVIDES A VIEW of the true character of the first 25 miles of the Willamette River. Relatively large segments of cottonwood-lined riverside can be found, with significant floodplains along the river. Agriculture makes its presence felt as well, especially in the production of grass seed for lawns, mainly Kentucky bluegrass, ryegrass, orchardgrass, and fescue. According to the Oregon Department of Agriculture, there are more than 1500 grass seed farms in Oregon. Some 60 percent of the world's grass seed comes from Oregon, and 95 percent of that grass seed is grown in the Willamette Valley. Grass seed is typically harvested in late June and early July. After harvest, fields were traditionally burned. In recent years, though, people have been voicing concern over the impact of field burning on air quality and its effect on residents' respiratory health.

Many other, more traditional crops are grown in the Willamette Valley as well, although less so next to the mainstem river. With a greater emphasis on local food production, more small farms are coming onto the scene to meet the increasing demand for tasty, locally produced fruits and vegetables. This is no surprise, given that many people migrated to the Willamette Valley during the nineteenth century to farm the fertile soils found here. Early on, men freed from their contracts with the Hudson Bay Company at the mouth of the Columbia eyed the valley with a sense of possibility. Perhaps they could develop prosperous farms in this mild, cloud-covered landscape? Early missions, such as the Methodist Episcopal mission of Jason Lee established north of Salem in 1834, brought with them the idea that Americans could settle the valley and take advantage of the breadth of fertile lowlands. After 1843, when the Provisional Government of Oregon was established, increasing numbers of white settlers arrived in the Willamette Valley, in part motivated by the 1839 bill that asserted U.S. ownership of Oregon and promised 1000 acres of land to every white male over 18 years of age who settled in the territory. The rich

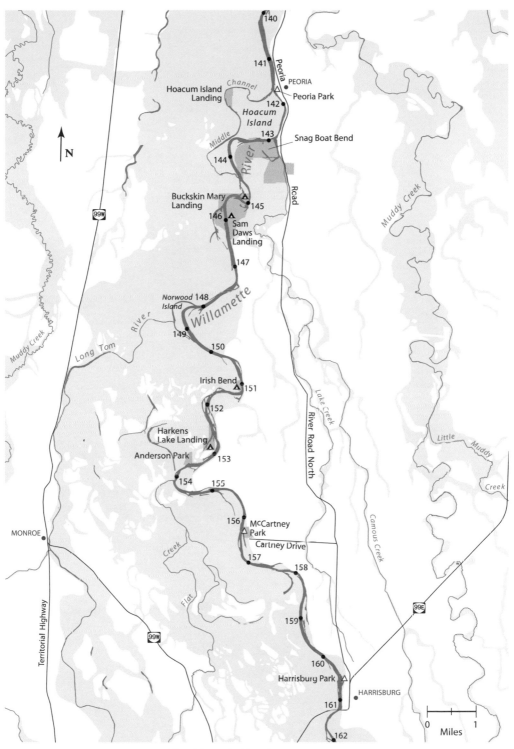

140

141

Peoria

PEORIA

Hoacum Island
Landing

Channel

Peoria Park

142

*Hoacum
Island*

143

Snag Boat Bend

144

Middle

River

Buckskin Mary
Landing

145

146

Sam
Daws
Landing

Road

147

*Norwood
Island*

148

Willamette

149

150

River

Irish Bend

151

Long Tom

152

Muddy Creek

Harkens
Lake Landing

Anderson Park

153

River Road North

Lake Creek

Little

Muddy

154

155

Creek

Camous Creek

MONROE

156

McCartney
Park

Cartney Drive

157

158

99W

Creek

99E

159

Flat

160

Territorial Highway

Harrisburg Park

HARRISBURG

161

0   Miles   1

99W

162

Harrisburg to Peoria

Looking from Buckskin Mary Landing, field burning can be seen beyond Sam Daws Landing.

Willamette Valley became a destination point for thousands of settlers each year, with farms rising from the floodplain areas along the river.

This settlement pattern created a longstanding relationship between agricultural interests and the Willamette River that can be seen clearly to this day. Although agriculture provides many obvious benefits for people, depending on the operation and the landowner, agriculture has had some negative impacts on the river, including contamination from chemical runoff. Intensive agriculture mainly takes place along the Willamette's rich floodplains, and the same is true of many of its tributaries, such as the Santiam and the Calapooia Rivers. The transformation of these historic floodplains has greatly affected the Willamette's ability to function as a productive and healthy river system, as minerals and nutrients from the land can no longer enter the system.

From the air, one can easily see the floodplains along the Willamette, the expanse of low-lying land stretching outward. Under natural conditions, when rivers are running high—typically in late winter or spring—these low-lying floodplains are inundated with water to one degree or another. On this stretch of river, Peoria Road from Harrisburg to Corvallis runs across a vast floodplain. As you travel, the land that is so shaped and molded by the river is easy to see, gently rising and falling along the river and often only a few feet above the Willamette's average spring flow.

A typical farm in the Willamette floodplain

Peoria Road between Corvallis and Harrisburg

WHERE: From Corvallis, take Highway 34 east for 2 miles to Peoria Road, turn right and head south to Harrisburg. From Harrisburg, Highway 99 (3rd Street) forks to the left and becomes Peoria Road.

CONDITIONS: A two-lane highway.

AMENITIES: There are no services between Harrisburg and Corvallis.

Peoria Road winds its way across the lowlands near the Willamette River for nearly 15 miles. Along the way you can see the river periodically as it moves northward. From a vehicle, you can easily see the large expanse of land on both the east and west sides of the river.

In the relatively small flood of January 2006, the river expanded across some of this area, with the brown flow carrying large snags and other woody debris. Eventually this abundance of wood was removed by landowners. At Peoria County Park just north of Peoria, near the parking lot you can look across the river to where the landowner has installed stream barbs, piles of large rock that act to divert the flow of the river back into the main current, deflecting the erosive power of the river away from the riverbank.

A mile past Peoria, you can visit a section of Snag Boat Bend Wildlife Refuge, part of the U.S. Fish and Wildlife Service's William L. Finley complex. While this refuge is not vast, it does provide excellent habitat for a range of

Stream barbs are constructed to deflect the water's force from the riverbank, thereby protecting it from erosion. While this is likely better for wildlife than rip-rap, other more natural methods can be used.

species that call the Willamette River home. There is a good parking lot just off Peoria Road (turn right at the sign). From the parking lot, take the path northward to the extensive backwater that spans the refuge. Through binoculars, here you can often see western pond turtles (*Clemmys marmorata*) basking on the exposed logs and snags. The western pond turtle, also known as the Pacific pond turtle or Pacific mud turtle, is one of only two freshwater turtles native to the Pacific Coast west of the Sierra-Cascade divide. The area is also frequented by green herons (*Butorides virescens*), great blue herons (*Ardea herodias*), great egrets (*Ardea alba*), myriad swallows, and other birds.

Heading southward from the refuge on Peoria Road, again take in the view of the vast Willamette floodplain. Here it is easy to imagine the impact of high water on the landscape, historically spreading for miles in every direction. Also clearly visible in this area is the dominant use of the Willamette lowlands for various types of agriculture.

From Peoria Road, Cartney Drive leads to McCartney Park in Linn County, where a small boat ramp provides ready access to the Willamette. The Harkens Lake greenway site (RM 153) can also be accessed by river from this park. In addition, as you drive keep an eye out for northern harriers (*Circus cyaneus*). These large, long-tailed, gray (male) or mottled brown (female) hawks fly low over the fields searching for prey. The last stop on Peoria Road is Harrisburg, which has a nice riverfront park with a large gravel bar and boat ramp ■

Herons

Throughout the year, great blue herons (*Ardea herodias*) can be seen standing still along the river, looking down into the water for fish, or high up in overhanging branches. These majestic birds stand 4 feet tall and have long yellow beaks and a white crown stripe above a black band. Their chests and long S-shaped necks have short gray feathers. Herons have an almost primitive quality when they fly, with their necks somewhat retracted, pumping their great wings as they glide between the cottonwoods. If you get too close a great blue heron will take flight, often letting out a series of croaks. Heron nesting areas can be seen all along the Willamette. These rookeries are typically high up in the cottonwoods and consist of several to many large nests

Great blue herons often stand still at the water's edge, waiting for wayward fish.

A great blue heron rookery perched high in the cotton-woods. Such rookeries can be seen all along the river.

clustered together. The nests are an assemblage of large sticks. Most eggs are laid in March, and it takes about 28 days for them to incubate. If you're lucky, it is possible to hear these birds near the rookeries in the spring as the young hatch and a constant chatter ensues.

The green heron (*Butorides virescens*) is common along the Willamette River. This wading bird is 16 to 18 inches tall, only about a third the size of the great blue heron. The green heron's plumage and bill are dark, and it often keeps its long neck pulled in close to its body. The green heron is typically seen perched among dense riparian vegetation, though it can be difficult to see as it stands motionless in the water waiting for a fish to approach. The green heron is one of the few birds to use tools. It drops bait, including worms, insects, twigs, or feathers, onto the surface of the water and catches the small fish that are attracted. This species also has a knack for zipping upward and away as you draw near in a boat or on foot, so getting a close look at a green heron is not easy.

Harrisburg to Peoria, RM 161 to 141.5

STARTING POINT: Harrisburg Park in downtown Harrisburg, at the boat ramp or adjacent gravel bar

ENDING POINT: Peoria Park, just off Peoria Road approximately 15 miles north of Harrisburg

DISTANCE: 20.5 miles

SKILL LEVEL: Experience with strong current and woody debris.

CONDITIONS AND EQUIPMENT: This stretch works well for canoes, kayaks, and drift boats. There is occasional woody debris and strong current.

AMENITIES: Harrisburg Park has free parking, restrooms, and running water. Peoria Park has free parking and pit toilets.

WHY THIS TRIP? You can get a great view of Marys Peak to the west and you pass Norwood Island, Sam Daws Landing, and Buckskin Mary Landing, as well as the Snag Boat Bend unit of the William L. Finely Refuge.

You catch the current quickly when putting in at the Harrisburg boat ramp or the adjacent gravel bar, where the river is a few feet deep. The river bends to the left and then right again as you pass the large gravel bar separating the Harrisburg sewage lagoon from the river. (The lagoon sits well above the river.) The river divides downriver near RM 159. This type of area can change seasonally, so caution is always advised for the river traveler. The small island owned by the Department of State Lands sometimes has a small backwater and fast shallow flow at the river's edge. Along the way you'll notice stands of cottonwood on private properties, nursery trees at the river's edge, and ample evidence of surrounding agricultural operations.

Although very pressing habitat issues are visible all along this stretch of river, there are promising areas as well, such as Harkens Lake Landing (RM 153). This Oregon Parks and Recreation Department greenway park is situated along an inside bend of the river, and upstream is a small undeveloped park owned by Benton County. Harkens Lake gains its name from the backwater area that stretches inland nearly a half mile, roughly parallel to the river. A few logs and roots dot the shallow entrance to the backwater, which is rich with native plants along the riparian fringe, including abundant red osier dogwood (*Cornus sericea*) and, nestling at the water's edge, Pacific willow (*Salix lucida* ssp. *lasiandra*). Separating this backwater from the main channel is a large peninsular formation with a forest of black cottonwood (*Populus balsamifera* ssp. *trichocarpa*), Oregon ash (*Fraxinus latifolia*), and other trees and shrubs. The gravel bar area is a Willamette Water Trail campsite, as well as the mainland area just inside the backwater.

The gravel bar area is very scenic, overlooking the eastern bank with a steep drop. At the water's edge are numerous trees, some nearly ready to fall into the river with tangled roots extending into the air in some cases. In the woods behind the gravel bar, deer can be seen and some people have even reported seeing Roosevelt elk (*Cervus canadensis* ssp. *roosevelti*), which likely traveled from the William L. Finley National Wildlife Refuge. This fantastic wildlife refuge, operated by the U.S. Fish and Wildlife Service, is a few miles west of this greenway site.

What makes Harkens Lake special is that it has been left alone. While locals sometimes ride their four-wheelers along the rocky shoreline, a practice that is *not allowed* on Willamette greenway properties, the site is usually quiet. Take a half hour to explore the backwater and feel the water, for at times it may be cooler than the water in the mainstem Willamette. In the summer months, large bryozoans can be seen floating near the surface of the backwater area. These algae-like masses are actually members of the animal kingdom and with a little imagination seem like greenish brain matter.

Norwood Island is located at RM 149, at the confluence of the Long Tom and Willamette Rivers. This 80-acre flat island was farmed as recently as the mid-1990s. On the west side of Norwood Island is a shallow backchannel that contains a wonderful bed of freshwater mussels, with the lower half full of current due to the flow of the Long Tom River from the west.

Sam Daws Landing and Buckskin Mary Landing can be found at RM 146 to 145. These two greenway parks are immediately adjacent to each other and are large gravel bars covered with willow and cottonwood. Once a peninsula extending from the west, Buckskin Mary Landing is today an island. In 2005–

An old car was dumped into the river across from Norwood Island.

2006, the main channel of the Willamette punched through a small channel. The resulting erosive force of the river created a much larger channel and took some land from the west riverbank. This area is ripe for a creative fix that could help secure the bank and provide some natural attributes for the river, such as large woody debris and occasional large rocks. A golden eagle (*Aquila chrysaetos*) was reportedly sighted here. These large eagles are more frequently found far inland and are only occasionally spotted in the Willamette Basin. Golden eagles look very similar to juvenile bald eagles, but they have been tracked through the Willamette Valley on their long migrations into Canada, so perhaps that great bird was indeed here. You can spend a morning trekking around Sam Daws Landing, a gravel bar replete with a backwater area, willows and cottonwoods, and a nice array of birds in the spring. If you camp there at night, the sound of the nearby Pope & Talbot Mill in Halsey several miles east can be heard, yet the area feels remote.

Just downstream from Buckskin Mary Landing is the opening to Hoacum Channel, also called the Middle Channel, delineating a large floodplain island at RM 144.5. At times, this channel has seen vastly reduced flows from large collections of snags and other woody debris that flowed in. When conditions are right, this 2-mile channel can be paddled, eventually reaching Hoacum Island Landing, an Oregon Parks and Recreation Department greenway site. Much of the side channel is riprapped, but initially there is an abundance of robust riparian vegetation. You may notice piles of rock along the shoreline of the river, known as stream barbs. These collections of rock are an improvement over riprap, especially if there is a robust riparian planting between them.

The Snag Boat Bend unit of the Finely Refuge is located at RM 144 to 143. Snag Boat is a 341-acre floodplain area with a significant backwater. Here you can easily see western pond turtles during the warmer months, usually basking on logs, and numerous birds and mammals as well. While traffic sometimes can be heard from Peoria Road, paddling up the backchannel can be well worth the time. In this stretch, the river maintains a variety of currents, with swirling eddy lines here and there, transitioning to fast water on the outside bend across from Peoria. Houses can be seen on the final half mile into Peoria. You will not see the boat ramp from the main channel, as it has been separated from the main flow by a large gravel bar. Just keep a close eye near where the Hoacum Island channel enters the mainstem. ▣

In an effort to increase the riverside habitat suitable for wildlife, state and federal agencies have been working with landowners to develop innovative new ways to reduce erosion. While this can be difficult with a river the size of the Willamette, it has not stopped people from trying. Several projects along the Willamette in Benton County show that dumping hundreds of metric tons of rock is not the only way to protect riverside areas.

Unfortunately, not all landowners or government entities have accepted new ways of relating to the river. At RM 151, a recent bank stabilization project was conducted in the old way of dealing with the river. On the west side is a Benton County Park known as Irish Bend, on the east side is a steep bank and Linn County property. Significant erosion of the steep bank occurred with the high water in late December 2005 and early January 2006, and the edge of the road in Linn County began to fall into the river. As the flood receded, Linn County assessed the damage and decided to seek funds from the Federal Highway Administration to fix their road. Linn Country might have chosen to move the road away from the river but instead decided to armor the Willamette's bank with large rock. Before most people concerned with the river were aware of the county's actions, including Benton County

The old ferry at Irish Bend (c. 1930s). Benton County Historical Museum & Society, Philomath, Oregon, no. 20000260061

Parks who own the park across from this site, Linn County went forward with large equipment and scaled the bank of the Willamette backward, making it more gradual. They then placed large rock in the river at the base and worked up toward the road, eventually armoring the entire bank for nearly 200 yards.

Across the river at Irish Bend Park in Benton County, you can see the old cut bank that was riprapped.

In a time when the U.S. Army Corps, who helped or led the riprapping of most of the Willamette, is beginning to think about how to restore American rivers, this county project conducted with federal funds moved the idea of restoration backward. Instead of fully armoring the bank, Linn County could have scaled it back with a more gradual slope, used some rock and collections of rock at the base, and then vigorously planted the riverbank with a variety of native species. Unfortunately, this type of approach was never considered.

As a river turns, much of its flow is directed against an outside bend. In this case, the outside bank of the river is now hardened, thereby pushing the flow back and toward the park on the opposite side of the river. Just downstream of Irish Bend Park on the west bank is some very low-lying farmland. What might increased flow bouncing off the upstream riprap do to this farmland? Too often such questions have not been asked during the planning of projects that manipulate the natural structure and function of a river. The state and federal natural resource agencies may well ask Linn County to vastly alter their project or conduct mitigation elsewhere. In addition to the impact on the flow of the river, the project provides no useful fish habitat and nothing in the way of riparian vegetation, which are two important considerations for the design of any such project. Although we can sympathize with the loss of a part of a loop road that serves a few residences, forcing them to drive one way in and one way out, that was no excuse to put aside all that has been learned about river function and revert back to the old way of doing things.

In a more positive vein, there are many good restoration opportunities along the river, with an increasing number of landowners who want to do the

Sterile and ill planned, this riprap project in Linn County does little for the river. Other options that work better for people and wildlife were available.

A young resto-
ration planting
on the main
channel side of
Norwood island.
The plastic
sleeve helps to
keep deer from
grazing on the
growing plant.

right thing for the Willamette. Norwood Island at the confluence of the Long
Tom and Willamette Rivers is one good example of this growing trend. The
Long Tom is a slow, meandering river whose headwaters are southwest of
Eugene. Water in the river is controlled by releases from the Fern Ridge res-
ervoir, a popular recreational spot. Snaking to the north and east, the Long
Tom meanders some 30 miles through farmland before it reaches the Willam-
ette River.

   After the flood of 1996, the stout bridge on the backchannel of Norwood
Island was damaged, and the landowner deemed the value of farming on the
island to be less than repairing the bridge. The landowner also understood the
need for restoration along the river and was very agreeable to this sort of activity on the property. Working with Benton County and a restoration organization called Cascade Pacific, RC&D, the landowner entered into a conservation easement. Under an easement, a private landowner agrees with a third
party to discontinue active agricultural use or development and leave the
property in its natural state or allow restoration activity to occur. The type of
activity allowed is recorded in the deed, thereby protecting the property for
future generations. It is hoped that more landowners along the Willamette
will agree to conservation easements.

   To gain clean water and healthy habitat, it is important to establish the
best restoration approach, defining what will yield the most ecological benefit
as well as the overall feasibility and cost of a project. Norwood Island was
already planted with native trees, providing some wildlife habitat, and has a
relatively productive backchannel. One option for improvement is to further
protect the mainstem bank of the island with native plantings and woody

debris, which can prevent erosion and improve habitat along the riverside. Organizations such as Willamette Riverkeeper and various watershed councils will make an effort to restore Norwood Island, and it would be an excellent candidate for public ownership as well, perhaps as an Oregon Parks and Recreation Department greenway site.

Freshwater Mussels

Mussels are a mainstay in the Willamette River, which is home to several species. The western pearlshell (*Margaritifera falcata*) and Oregon floater (*Anodonta oregonensis*) are the most common. These amazing mollusks, with a bivalved oval shell, can live longer than most other freshwater species. For instance, the western pearlshell can live for more than 100 years. Pearlshell mussels are often found in relatively shallow side channels, with clear moving water, such as the backchannel of Norwood Island. Mussel shells can be seen occasionally littering riverside areas where predators, such as river otters, are able to get at them.

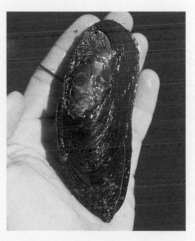

A western pearlshell mussel from the Norwood Island backchannel

Freshwater mussels have an interesting life cycle. After adult mussels breed, the embryos develop into larvae (called *glochidia*) within the female mussel's shell and are subsequently released into the water. Each larva then attaches to a host fish, which transports it for weeks until it drops off and lands on the river bottom. The tiny growing mussel then burrows into the sediment and stays buried until it reaches maturity. A mussel will stay in the same general area once mature. Mussels can be found in beds of hundreds of individuals.

In 2006 and 2007, Willamette Riverkeeper conducted a study of mussel populations in the Willamette system. More than other species, the western pearlshell commonly relies on cold, clear streams. The study provided important data about the location and abundance of western pearlshell mussels along the river and shed light on key areas where adequate habitat still exists for this species. Because of their long lifespan and sedentary life habits, pearlshells can tell us a lot about the health of our river systems over time. There is also growing concern about the fate of the pearlshell as it relates to the host fishes that transport the larvae. If native fish species, such as cutthroat trout, are reduced, then populations of western pearlshell and other mussels will be affected as well.

With so much land along the Willamette and throughout the Willamette Valley having been modified over the years, habitat restoration is one of the highest priorities that exist today among state and federal natural resource agencies, nonprofit organizations such as Willamette Riverkeeper, local watershed councils, and academic institutions such as Oregon State University. Although bringing the river back to its original state is a laudable goal, the more resonant and pragmatic need is to bring the river back to some semblance of its former state. Decades of habitat modification make this an enormous task, with many years and the commitment of numerous agencies required to get the river to a point where it can support even a modest number of self-sustaining native species. Improving the Willamette River will entail improving natural ecosystem function through habitat restoration and changes in the management of hydropower dams, controlling wastewater discharges from pipes and runoff of contaminated water, and cleaning up contaminated sites, such as riverside dumps, and contaminated sediment in the river itself.

This cut bank along the Willamette is a good example of an area that would benefit from riverside plantings and perhaps sloping the bank. Such projects can help curb erosion, a problem for landowners, and provide improved riverside habitat for wildlife.

Of the river's native fish, spring chinook (*Oncorhynchus tshawytscha*) and Oregon chub (*Oregonichthys crameri*) are listed as threatened and winter steelhead (*Oncorhynchus mykiss*) is affected as well, with their former naturally reproducing populations greatly depleted. This depletion is related, in part, to the presence of large dams on the Willamette tributaries, which largely prevent the fish from returning to their spawning grounds, as well the overwhelming modification of riverside lands. When a river can no longer sup-

port species that have thrived there for thousands of years, there is a clear problem. The overall reduction of the river's complexity is emblematic of what has happened to the Willamette and its tributaries. Human activity has eliminated the many side channels and alcoves that provided rich, cold-water habitat that helps to sustain these fish species. If the Willamette has any chance of sustaining native species, habitat restoration must be a focal point for the next two decades and beyond. There are several basic ways, across a variety of scales, that habitat can be restored.

Riparian restoration is a relatively simple way to make improvements, which involves replanting those riverbanks that have been stripped of native vegetation as well as placing large root wads and logs to mimic a natural floodplain. Over time, such planting produces shade, thereby helping to keep the river water cool. This is especially important on the smaller tributaries.

Side-channel restoration is a more complicated action that opens up remnant parallel channels of the river to fish and other wildlife. Before they were cut off from the main river's flows, such channels may have contained water year-round or only seasonally. This form of restoration often depends on landowners willing to provide the necessary acreage for this use. Side channels can be in floodplain areas, as well as within the bounds of the existing river channel.

Floodplain restoration, while related to side channel restoration, occurs when low-lying areas along the river are opened up to high flows. Such areas may not have a defined exit back to the river downstream, meaning that the water can inundate low-lying areas for a while before retreating back the way it came in. Most often, such areas have been separated from the main river by human-made structures designed to keep water from expanding outward onto the lowlands adjacent to the river. When inundated by high water, floodplains provide a rich assemblage of nutrients for fish, as well as areas to take refuge.

If riparian, side-channel, and floodplain restoration were to take place on a wide enough scale, along with improving fish passage at the dams within the Willamette system, we could well see the recovery of spring chinook within our lifetimes. There is no way to accurately predict how quickly such recovery would occur, but there is a clear relationship between natural functioning of a river system and healthy native fish populations. The challenge for the Willamette River, however, is to enable this ecosystem function to occur amid a river basin that has significant logging in forested headwaters with multiple dams, within a wide valley harboring hundreds of thousands of acres of agricultural land, millions of people, and several major cities along the river's banks.

There are some excellent examples of habitat restoration occurring along the Willamette, with the needs of both landowners and the river being met through creative, nontraditional solutions. At RM 147, the landowners decided to join in a project to protect the bank without using any riprap. Every year the owners were losing land because the relatively steep bank was being eroded and dirt and rock were falling into the river at high flows. With the cooperation of a local restoration agency, the owners sculpted the steep bank backward, thereby making the overall slope much more gradual. Next, large logs were secured into the bank with cables to ensure that they would not wash out at high flows. These assemblages of logs help divert some flow back to the river, while creating an eddy behind them at high water. Between the log collections, the bank was planted with native vegetation. So far, this has seemed to work. Although some of the plantings died—as some level of plant mortality is expected in all riparian restoration projects—the logs remained in place. This was tested in the winter of 2006, when high flows inundated the area. While no one can predict how long a riverbank will remain in place, there are encouraging signs that more natural bank protection methods can work for the long term and provide meaningful improvements to habitat.

American Beaver

A key wildlife species along the Willamette, the beaver (*Castor canadensis*) is the largest of the rodents in North America. It is closely linked to the identity of Oregon, which is known as the Beaver State. Beginning in the early 1800s, trappers sought this animal throughout the Northwest. The Hudson Bay Company staked its livelihood on a large population of beavers in the area, as trappers made their way from one river system to another, taking as many animals as they could and bringing back the valuable pelts to the company's headquarters at Fort Vancouver. Although the beaver serves as mascot for Oregon State University's sports teams, this amazing wild animal is a mere afterthought for most people. Yet beavers are relatively numerous along the Willamette River and its tributaries.

A beaver shares the beach with a wood duck.

Beavers have dark brown to reddish brown fur, with coarse hair on the outer layer and a finer layer underneath. They gain their waterproofing from a gland that secretes oil from their cloaca, which they spread to coat their fur. The scented secretion castoreum is used to mark territory. Another amazing trait is the nictitating membrane that covers their eyes when they are underwater, allowing them to see. To stay warm underwater, beavers rely on an ample amount of fat for insulation. Their large heads have a triangular appearance when they peek above the surface of the water. Beavers usually mate for life, and their young (known as kits) stay with the parents for up to 2 years. They can grow up to 65 pounds and 4 feet in length. Nutria, a nonnative species, can sometimes be confused with the beaver, but the nutria is smaller and has a long thin tail, as opposed to the large paddle shape of the beaver tail.

Beavers are usually active at night, but along the Willamette they may sometimes be seen during the daylight hours. Swimming is second nature for beavers, and they can stay below for nearly 15 minutes, though most often you will see them dive and then return a short time later. When a threat arrives nearby, beavers will often slap their large flat tails loudly, marking their presence and territory.

Beavers' classic log-and-stick lodges not only back up water (and thus are commonly called dams), but provide a relatively safe and warm home for the animals. Beaver dams also provide benefits for fish by backing up water in key areas that can provide refuge from high winter and spring flows. These ponds can also create habitat for various waterfowl. In recent years, the ecological contributions of beavers have been more widely acknowledged by fisheries biologists and others. Along the mainstem Willamette you will more often see beavers with riverside dens, usually dug in high banks, underneath the brush. There are often slides, where it is evident that a good-sized furry creature slides down the dirt bank into the river. In fact, along many rivers in western Oregon, beavers utilize bank dens.

A telltale sign of beaver is the presence of light-colored sticks peeled of their bark and floating in the water or sitting on shore. You may also see larger trees that have been cut down. Beavers patiently chew away at trees with their large, long teeth, eventually working them down to a small point in the middle, which causes the tree to fall over. Beavers leave a characteristic cone-shaped cut when they fell a tree. People have developed clever ways to discourage beavers from cutting down trees in private yards and parks. A combination of wire mesh along the tree trunk and spicy coatings have worked in many instances. With some creativity, sensibility, and compassion for wildlife, people can coexist with wildlife such as beaver and be the better for it.

Beavers chewed these trees down during the peak of a flood event in 2006. You can see their efforts up and down the trunk.

Just downriver of this property is the wastewater discharge for the Pope & Talbot Mill at Halsey, which produces pulp for paper production. In the late 1960s, this mill was a test case as it sought a permit to discharge wastewater into the river after new rules had been enacted by the state of Oregon regarding water quality in the Willamette. Ultimately, the Oregon Department of Environmental Quality was convinced that the mill represented a new breed that would not significantly impact the river. The Environmental Quality Commission permitted the mill. Though this facility was an improvement, one would be hard-pressed to call the discharge clean. The residents of Corvallis have concern over this, given that their drinking water is drawn downstream of the mill.

Along this stretch of river from Harrisburg to Peoria, it is also common to see birds of prey, including bald eagle (*Haliaeetus leucocephalus*), Cooper's hawk (*Accipiter cooperii*), and merlin (*Falco columbarius*). The bald eagle, our national symbol, is easily recognized by its brown body and white head and tail. This plumage coloration, however, is attained at maturity, around 5 years of age. Immature bald eagles vary in color from all brown to mottled brown and white, depending on the bird's age. The Cooper's hawk is a medium-sized hunter of the forest that specializes in eating birds. With a long rounded tail and short rounded wings, this hawk is built for fast flight through the obstacle course of trees and limbs. The Cooper's hawk has a dark gray or grayish brown back, and its underparts are barred reddish and white. In contrast, the merlin is a small falcon with long pointed wings and a long banded tail. Its head has a faint mustache mark, the chest and belly are streaked brown, and the back is gray or brown. Merlins do not build nests, but instead take over old nests of other raptors or crows.

The backwater area on both the west and east sides of the river can harbor a variety of shoreline birds, from great egret (*Ardea alba*) and great blue heron to spotted sandpiper (*Actitis macularius*) and greater yellowlegs (*Tringa melanoleuca*). The great egret is found across much of the world, from southern Canada to Argentina and in Europe, Africa, Asia, and Australia. This large (nearly 4 feet tall), all-white heron has long black legs and feet, and its yellow bill is long, stout, and straight. The spotted sandpiper breeds along the edges of a wide variety of water sources, from the backwaters of the Willamette to urban ponds, and gets its name from the distinct round spots on its belly. The greater yellowlegs frequently announces its presence by piercing alarm calls. This tall, long-legged bird lives along freshwater ponds and tidal marshes and is mottled brownish, with a white rump and tail and bright yellow legs.

Natural areas such as McCartney Park, Buckskin Mary Landing, and Sam Daws Landing provide a great opportunity to experience floodplain lands. You might notice that some of the Oregon Parks and Recreation Department properties could use some work to remove invasive species, such as Scotch

Killdeer

Killdeer (*Charadrius vociferous*) are common along the Willamette River, in open fields, and even in roadside areas. These birds definitely enjoy the Willamette's gravel bars and backwater areas, where you can see them standing along the shoreline in the rocks or exposed sand. Killdeer are about 10 inches long with a wingspan up to 24 inches; they are larger than sandpipers, which also occupy similar riverside areas. The birds have a large, distinct black band around the base of their necks.

Killdeer are well known for their persistent, piping, and drawn-out calls that seem to erupt whenever one feels threatened or when the well-camouflaged eggs are encroached upon. Killdeer fly away from their exposed nests, seeking to draw potential predators away. It is common to see them throughout the year, although they are most often seen along the Willamette in the spring and summer months, when they are most active. If you make camp or stay at a riverside area near their nests, be prepared to listen to them making their displeasure at your presence vividly clear.

Killdeer can be seen along gravel bars and exposed mudflats on the Willamette.

broom (*Cytisus scoparius*) and Himalayan blackberry (*Rubus armeniacus*), but they also provide a good amount of native vegetation, especially willows and cottonwoods. The federal wildlife refuges in the Willamette Valley also play a vital role in providing species a place to flourish, though they are relatively few and far between. At just over 5000 acres, the William L. Finley National Wildlife Refuge is a U.S. Fish and Wildlife Service property located a couple miles west of this stretch of river, off Highway 99W. It provides a wide sweep of habitat that can be essential to the health of birds that migrate through, and the refuge also sustains resident species such as the Roosevelt elk. The 341-acre Snag Boat Bend unit at RM 143 is part of the Finley refuge.

4

Peoria to Albany
RM 141 to 120

*What we should be united against is the needless
destruction and pollution of the world we live in.*
—BILL MASON

THE STRETCH OF RIVER from Peoria to Albany is fairly scenic, but it has a bit of everything: large tracts of agricultural land growing a variety of crops, industrial facilities, municipal waste treatment facilities, and large greenway parks. This section provides a good sampling of human interaction with the river.

Peoria is a small town approximately 10 miles south of Highway 34 on the east side of the river. The town was once more bustling than it is today, but at present Peoria is simply a collection of homes along the river. There are no stores or services in town. The boat ramp at Peoria Park is behind a large gravel bar, and over the years the bar has grown, making access with powerboats during low water a bit difficult. Like many other gravel bars along the Willamette, this one is covered with Pacific willow. Paddle craft can access the ramp year-round, in sight of a nice patch of wapato (*Sagittaria latifolia*), or Indian potato, that grows along the edge of the gravel bar.

Downstream from Peoria at RM 140 is the opening for the main side channel that separates John Smith Island and Kiger Island. The smaller Clark Slough opens about a half mile upstream from this channel, but it is typically

The old Peoria Ferry in 1927. Benton County Historical Museum & Society, Philomath, Oregon, no. 20000260060

impassable. The two islands represent the braided channels of the historic Willamette River. In former days, these large floodplain islands would have been inundated with water on a regular basis and their channels morphed from year to year. Even in recent years, with tight control of river flows by the dams, high water has brought change to these channels.

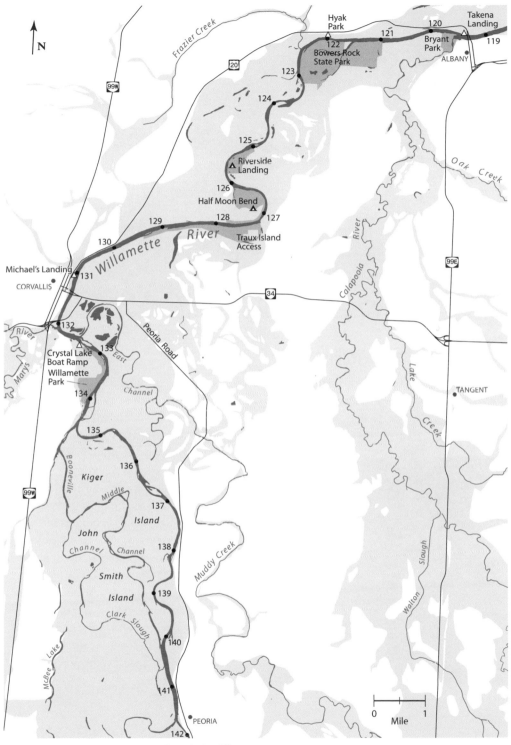

N

Frazier Creek

Hyak
Park

Takena
Landing

120

Bowers Rock
State Park

121

119

122

Bryant
Park

ALBANY

20

123

Oak Creek

124

River

125

Riverside
Landing

126

Half Moon Bend

Calapooia

127

Willamette   River

128

129

Traux Island
Access

130

34

99E

Michael's Landing

131

CORVALLIS

River

132

TANGENT

Marys

East

133

Crystal Lake
Boat Ramp
Willamette
Park

Peoria Road

134

Channel

135

Lake

Booneville

Kiger

136

Creek

99W

Middle

137

Island

John

138

Channel

Channel

Muddy Creek

Smith

139

Island

Clark

Slough

140

Walton

Slough

McBee Lake

141

0   1

Mile

PEORIA

142

Peoria to Albany

The practice of cutting off side channels has been commonplace along the Willamette. At a site a few miles upriver of Corvallis, this boat drove pilings into the river bottom at a side channel opening, thereby diverting the river's flow to the main channel (c. 1900). Benton County Historical Museum & Society, Philomath, Oregon, no. 2000-078.0008

Pacific Willow

The elongated leaves of the Pacific willow are distinctive.

Pacific willow (*Salix lucida* ssp. *lasiandra*) can be found dotting gravel bars and riverside areas along much of the Willamette River. Willows need sunlight, so they depend on disturbed areas that get flooded often, and gravel bars fit these needs precisely. Along the Willamette you can find younger stands of willows from waist high to ones that extend well over your head, all maintaining a brushy aspect. Larger mature willows can be found in groves on low river benches next to the river. Most commonly, you'll see Pacific willows extending the breadth of a gravel bar, and in summer months they are typically 20 to 40 feet from the water line. Wildlife can be found in their inviting thickets, and in the spring and it is common to see birds zipping along through the sea of willows on islands in the river.

Peoria to Corvallis, RM 141 to 132

STARTING POINT: Peoria Park, just off Peoria Road approximately 15 miles north of Harrisburg.

ENDING POINT: Just south of where Highway 99W crosses the Marys River in Corvallis, take Crystal Lake Drive east to the park entrance and the Crystal Lake boat ramp.

DISTANCE: 9 miles

SKILL LEVEL: Experience with strong current.

CONDITIONS AND EQUIPMENT: This stretch is good for canoes and kayaks. It is mostly flat water with a few areas of strong current and occasional strainers.

AMENITIES: Peoria Park has free parking and a pit toilet, and Crystal Lake has free parking and portable toilets.

WHY THIS TRIP? This is a good stretch for the whole family, with only occasional stretches of swift current. This stretch also gives a good sense of agricultural lands along the river and has a few small gravel islands owned by the Department of State Lands where paddlers can stop.

From Peoria the river moves at a good rate for much of the year. At RM 140 is the 8-mile-long channel that separates Kiger Island and John Smith Island. This channel can be very scenic in its upper stretches, with fast shallow water that affords wonderful views of local wildlife. If paddling this stretch, you must be able to easily control your craft in current and be very watchful of newly deposited logs and other woody debris. The lower portion of the side channel has little current.

Wandering along gravel bars can enable you to see evidence of wildlife, such as these deer tracks.

A river-view development at Corvallis

The side channel rejoins the river at RM 134.5, where there is a housing development near the river's edge. Soon you will see the main water intake for the city of Corvallis. The river rounds a long left turn and on the west side is Willamette Park, a sprawling city park that overlooks the river. The park is for day use and affords some beach access for local residents.

At RM 132.5 is the Crystal Lake Boat Ramp, owned by the city of Corvallis and immediately adjacent to Willamette Park. Here you have good access into Corvallis and a good-sized parking area that can be used for overnight stays with permission of the city. ▣

The Willamette River is the main drinking water source of Corvallis, supplying thousands of homes and businesses with their daily water needs. As you might imagine, it is important to the people of Corvallis that the water upstream is as clean as possible. Therefore, the National Pollutant Discharge Elimination System permits for the municipal treatment plants in Eugene and Springfield, as well as industrial permits for Pope & Talbot Mill in Halsey, are of great interest to the town's residents. Corvallis also has a good water treatment plant just a few blocks from the water intake. Ironically, wastewater is pumped back into the Willamette a couple of miles downstream.

The Corvallis area is rich with Willamette history. The old town of Orleans, which was situated on the east side of the river across from Corvallis,

The drinking water intake structure for the city of Corvallis

The ferry at Corvallis in 1907, with a great display of adults, kids, and livestock indicating how crucial this mode of transportation was along the Willamette. Benton County Historical Museum & Society, Philomath, Oregon, no. 1994-038.0667

was washed away in the flood of 1861. Just across the river from the Crystal Lake boat ramp in Corvallis is an old channel of the Willamette River, frequently referred to as the East Channel of the Willamette. River ecologist Patricia Benner has documented the history of this stretch of the river and found that this 2-mile curve of backwater was once the main channel. In 1906 the main channel diverted from the east, taking the more direct northward route seen today. This major change of direction serves as an excellent example of how the Willamette's course could change drastically, almost overnight.

The Corvallis waterfront after the flood of December 1890. S. B. Graham photograph, Benton County Historical Museum & Society, Philomath, Oregon, no. 1994-038.0284

The floodwaters of December 1890 near Corvallis. Benton County Historical Museum & Society, Philomath, Oregon, no. 2000-078.0006

The U.S. Army Corps of Engineers sought to protect the river from blasting through the peninsula by constructing a stabilizing wall, and pilings and rocks were placed along the riverbank of the former channel. Instead of pushing its way from the former bend across the lowland east of Corvallis, the river cut through the land on the upper end of a turn a couple miles upstream, which fixed the problem, making all the work on the outside bend of the channel for naught. Another of the Corps' solutions was to divert the river through today's channel, but the river provided that fix on its own. The remnant East Channel still exists today. The opening is just downstream from Crystal Lake boat ramp, and you can paddle up quite a way, traveling over old snags in relatively shallow water. The Corvallis stretch of the Willamette has also seen extensive work by the U.S. Army Corps to remove snags over the years.

Corvallis is also home to the Marys River, a tributary that meets the Willamette at RM 132. This little river has been improving ecologically for years, in large part thanks to the Marys River Watershed Council. This council was formed in the mid-1990s to restore and protect habitat along the river and is composed of stakeholders with a variety of perspectives. Local individuals with backgrounds in business, agriculture, academia, and conservation are involved in this watershed approach. In addition to improving habitat along the Marys River and the creeks that feed into it, the Marys River Watershed Council works to improve the ecological condition of upland areas. This work has aided a large breadth of habitat, benefiting not only fish and other river species, but also Oregon white oak (*Quercus garryana*), camas (*Camassia quamash*), and Fender's blue butterfly.

A view of the Marys River confluence with the Willamette (c. 1900). Benton County Historical Museum & Society, Philomath, Oregon, no. 2002-073.0026

Fender's Blue Butterfly

The endangered Fender's blue butterfly (*Icaricia icarioides* ssp. *fenderi*) lives primarily in the upland prairies of the Willamette Valley. The Fender's blue is a small butterfly, with a wingspan of about 1 inch. The upper wings of males are iridescent sky blue, and females have rusty brown wings. The upper surfaces of both sexes have a black border within a white fringe, and the underside is pearly gray to dirty chalk outlined in white, with black and brown spots.

Kincaid's lupine (*Lupinus sulphureus* var. *kincaidii*) is critical to this subspecies' survival. This wildflower is listed as a threatened species because of extensive habitat loss, and it is the primary food source for Fender's blue butterflies during the caterpillar stage. In May the butterflies deposit their eggs on Kincaid's lupines, and the eggs hatch in June. The caterpillars spend the winter in the root system of the plants. In March they emerge, crawl up the plants, and feed on the leaves. Once Fender's blues metamorphose into butterflies, they live for only a week to 10 days, during which time they mate and lay eggs.

Fender's blue butterfly was once thought to be extinct—the subspecies was known only from collections made between 1929 and 1937—until it was rediscovered in 1989. Historically, these butterflies thrived in the upland prairies of Oregon. Since the arrival of settlers in the state, however, an estimated 99 percent of this native prairie ecosystem has been destroyed. In January 2000, the U.S. Fish and Wildlife Service added Fender's blue butterfly to the Endangered Species List. Today, the Fender's blue occurs in thirty-two small sites totaling 408 acres, with twenty-four of these populations occupying sites of 8 acres or less. Fourteen sites are on federal, state, county, or city properties, and the rest are on private lands. Currently, the largest known populations exist in the Baskett Slough National Wildlife Refuge, west of Salem.

As an example of the strength that can be obtained with a solid, focused, experienced, and properly functioning council, in 2005 the Marys River Watershed Council brought together landowners in the Cardwell Hills area, just west of Corvallis, along the middle section of the Marys River. With assistance from the watershed council and other local partners, the landowners embarked on an extensive effort to improve wetland, riparian, and upland habitat that would improve the condition of oak savannah along the river, essentially restoring the entire watershed. In 2006 the state, through its Oregon Solutions Program, convened additional government representatives and nonprofit organizations to further assist the project. As of 2007, thirty-seven land-

owners in this section of the Marys River had joined in, with great enthusiasm. The Marys River Watershed Council represents one of the best councils in the Willamette Basin, with a strong degree of expertise and community support.

In the last decade, the city of Corvallis has embraced the Willamette by re-developing its waterfront. The city built walking paths, a community fountain, and benches and installed placards describing Corvallis history, including its long association with the river. In years past the river was of less importance for the downtown area, but today numerous businesses thrive there, with several facing the river. Michael's Landing, at RM 131, is just downstream of downtown Corvallis. This city park is undergoing some improvement to make river access better and provide additional amenities. This trend is occurring from Eugene to Portland, as towns large and small reorient themselves toward the Willamette River.

Stretching from Corvallis to Albany is a relatively wide, flat expanse of river that is fairly slow moving. The first few miles out of Corvallis parallel Highway 20, which runs along the west side of the river between the two cities. Four miles downstream of Corvallis, Half Moon Bend comes into view on the west side of the river at RM 127. This is a large undeveloped Oregon Parks and Recreation Department greenway. Slightly downstream of Half Moon

The Van Buren Bridge opens in Corvallis for the first time in 25 years as the U.S. Army Corps snag boat makes its way upstream in 1953. Benton County Historical Museum & Society, Philomath, Oregon, no. 19940380130

Perhaps this chair floating downstream of Corvallis was an engineering project for some inspired undergrad.

Bend is Riverside Landing between RM 126 and 125. Together, these two parks provide hundreds of acres of habitat along the Willamette and represent good opportunities for addition restoration work.

Just below these public lands, the river backs up slightly to go around an island. The larger channel is to the right, but the small left channel provides a good opportunity to experience shallow water making its way over a gravel bottom. The channels change from year to year, with room enough for a canoe or kayak some years, and very diffuse current in other years. Woody debris can be common in these shifting channels. Along such wide open areas of the river, keep an eye out for the occasional shorebird, such as great egret, great blue heron, spotted sandpiper, and greater yellowlegs.

This stretch of the Willamette is surrounded by agriculture. As in stretches upriver, irrigation water is taken from the river during the summer months when pipes and pumps can be heard humming away. Such diversions for irrigation typically take only a small portion of water from the Willamette. During periods of low flow, however, one may begin to wonder about the impact of irrigation and other water withdrawals from even a large river like the Willamette. Multiple water rights have been allocated for the Willamette, but not all of them have been used. If summer droughts begin to occur on a more regular basis in future years, perhaps due to global climate change, would the Willamette have the capacity to provide for all these water needs and still sustain a robust population of wildlife along its extent? As with many rivers across the world, water availability will likely become a more pressing issue in the future.

Greater Yellowlegs

Along the Willamette the greater yellowlegs (*Tringa melanoleuca*) is seen occasionally zipping along the shoreline and landing in backwater areas, where there is little current and ample opportunity to seek out food in the algae-rich shallows. Usually this bird is seen alone or with two or three others. At 14 inches long with a 28-inch wingspan, the greater yellowlegs is considerably larger than a killdeer. It resembles an extra-large sandpiper, with relatively long yellow legs and a long thin beak, but the greater yellowlegs has a characteristic white rump and tail and especially bright yellow legs. Seeing one of these birds along the Willamette is a memorable experience, given that they are wary of people and take off quickly when approached. Greater yellowlegs use the river most often in the summer months and are not known to nest along the mainstem.

Greater yellowlegs can be seen along the shoreline and in mudflat areas.

Bowers Rock State Park is another large public area along the Willamette from RM 123 to 121. This property has a sizeable backwater and is jointly managed by the city of Albany and the Oregon Parks and Recreation Department. Greenway parks like this are a sort of anchor natural resource. Some of these properties are in need of restoration work, while others need only be left alone. As of 2007, much of Bowers Rock State Park had poor access from land, but plans are in the works to provide improved access with opportunities for birding and other wildlife viewing. The park also provides a great opportunity for side-channel reconnection and other restoration work. The backchannel into Bowers Rock is very scenic and is worth a stop, as you might be treated to views of egrets, herons, and northern pintail ducks (*Anas acuta*). The northern pintail has a distinctive slim and long-necked silhouette. The male is easily identified by his white chest, white stripe up the neck, dark reddish brown head, and long tail, but even the dull-brown female can be recognized by her long neck.

The final few miles into Albany are bucolic. While this is not the most scenic or wild stretch of the Willamette overall, it does afford good opportunities to view wildlife. Here you can occasionally hear the hum of Highway 20

Willamette River, Albany, Ore.

The modern view of Albany has some resemblance to this photograph taken around 1905. Benton County Historical Museum & Society, Philomath, Oregon, no. 1994-008.0471

to the west, but by and large the area is quiet. Hyak Park (RM 122), a Benton County park just off Highway 20, has a parking area, restrooms, and a boat ramp. From here you can see across the river to Bowers Rock.

Bryant Park, at RM 120, makes a good put-in and take-out point and provides an interesting opportunity to view Albany's interface with the Willamette River. This park is located at the confluence of the Calapooia and Willamette Rivers. The 72-mile-long Calapooia meanders its way across a vast agricultural landscape from the Cascade foothills, flowing through Brownsville, then under Interstate 5, and snaking its way north and west until it meets the Willamette in Albany. Surrounded by a seemingly endless expanse of grass seed fields set against the small riparian fringe, the Calapooia River has retained enough healthy habitat to support native steelhead stock. In autumn of 2007 the Brownsville dam was removed, thereby creating access to some 17 miles of additional habitat for steelhead and other species. The dam removal action far upstream places renewed importance on habitat restoration in the lower portions of the Calapooia, to heighten the chance that fish can make is successfully upstream to spawn. Plans for the removal of another dam are in the works. This small Willamette tributary has relatively little flow through much of the year and thus could benefit greatly from riparian habitat restoration that creates cool water. The health of each and every Willamette River feeder river can make a difference to the health of native species.

North Fork of the Middle Fork of the Willamette River

Simple, affordable, and easy to store, a canoe can help you explore the Willamette.

A wonderful alcove just downstream of Eugene

As cities like Portland have grown and become more complex, the issues that affect the Willamette have done the same.

Endless time and water have shaped these river rocks along the North Santiam.

Seen from across Beacon Landing south of Eugene, a side channel that was formerly the main channel snakes its way across a sea of gravel.

A backwater area just downstream of Camas Swale on the Coast Fork

Camping along the shores of Waldo Lake can provide stunning sunsets.

Looking west, early morning sun bathes the lakeside rock of Waldo Lake.

Olallie Spring rushes out of the hillside, deep in the forest.

The intensely cold water of Great Spring is clear, with deep blues and greens.

As its name implies, Clear Lake's deep blue water is amazingly clear.

Scenic and lively in places, the Coast Fork is worth a look.

Ancient rock and moss on the flank of Mount Pisgah

Current rushes past the autumn colors on the Middle Fork.

The misty view from Mount Pisgah looking along the Coast Fork to where it joins the Middle Fork to form the mainstem Willamette

Winter morning light crosses the backchannel at Whitley Landing.

As it courses through Eugene and Springfield, the Willamette is interspersed by occasional rapids and fast-flowing water.

The confluence of the Willamette and McKenzie to the left, and the backchannel around Confluence Island rejoining the river to the right. Part of Green Island is in the foreground. Gravel pits can be seen in the distance, with a sizeable pit filled with water right at the McKenzie confluence.

Swift shallow water flows through ever-changing gravel bars near Beacon Landing.

High winter flows at Marshall Island Access.

Black cottonwood snags can be found all along the river.

This tree is just about to be recruited into the Willamette. Note the large exposed root wad.

A black cottonwood root wad at Blue Ruin Island

Blue Ruin visitors enjoy a faint rainbow.

The large gravel bar Blue Ruin Island holds fast, with most of the land in agricultural use on both sides of the river.

A quiet sunset on a gravel bar at Harrisburg

Harrisburg's Riverfront Park makes a great place to see the river and to get on the water. Here more than 100 people put in during a Paddle Oregon event.

Agriculture is the dominant land use along the river in this stretch, though the water treatment lagoons at Harrisburg (lower left) provide a change of scenery.

The quiet backwater at Harkens Lake Landing teems with wildlife.

The Long Tom River flows into the backchannel of Norwood Island.

Here a landowner sloped back an eroding bank, secured large logs at regular intervals, and planted the space in between with native vegetation.

A great egret searches for prey in a backwater.

Beavers have dammed up a backwater area along the Willamette.

Daisies push their way through the river rock at Sam Daws Landing.

A metal structure was built to help keep the bank from eroding, though in one winter large rocks were scoured out from behind it.

Willows typically dominate gravel bars along the Willamette and offer seclusion for wildlife.

Public spaces, such as this island owned by the Department of State Lands, make great campsites along the Willamette. Leave all campsites in better condition than you found them.

Corvallis provides some good access points to the river. Here participants in a Willamette Riverkeeper event take a break at Michael's Landing.

Flowing slowly from the Coast Range, the Marys River is a gem that has benefited from a host of ongoing restoration projects in the watershed.

Bowers Rock Park is a large public property that could provide excellent restoration opportunities.

Named after the Kalapuya people of the Willamette Valley, the Calapooia River snakes its way from the Cascade Range across a vast open plain. This nicely wooded spot is just a few miles east of Corvallis.

The scenery along the Luckiamute River is rich as this tributary enters Luckiamute State Natural Area.

The Luckiamute State Natural Area is the largest tract of lowland forest along the Willamette. Here you can see the surrounding agricultural land use and the Santiam River entering the Willamette at the lower right.

The Santiam River provides an abundance of wonderful scenery.

A rainbow over a small lake in the upland area of Luckiamute State Natural Area

The opening of the side channel at Wells Island

The scenic winding channel behind Wells Island

Two bald eagles perch along the river on a downed cottonwood snag.

These two photos of American Bottom Landing, which provides good camping along the river, show how flood-water can dramatically change riverside areas from one year to the next.

Independence Riverfront Park offers good access to the Willamette.

Looking from Wallace Marine Park toward Salem's waterfront and Minto-Brown Island

Providing a colorful contrast to the river rock, California poppies grow on many Willamette gravel bars.

Even near a city the size of Salem, riverside areas can be scenic.

Water rushes over a small gravel side channel at Keizer Rapids Park.

A window to the river framed by riparian trees

A cold winter sky reflected in the Wheatland Bar back-channel

Wheatland Bar and its backchannel are full of wonderful scenery.

The Wheatland backchannel flows through the river rock at the opening of Lambert Slough.

Native camas can bring an abundance of color to riverside meadows in spring.

Wapato occasionally can be seen along the river in large patches.

The Yamhill River enters the Willamette's Newberg Pool.

Smooth waves of gravel coat the upper end of Five Island.

A Canada goose on a high basalt face

Wide and flat, the Newberg Pool has relatively little current.

Paddlers enjoy the calm current near Newberg.

The Molalla joins the Willamette at Molalla River State Park.

Lively and cold, the Molalla River rushes from the Cascade Range.

Stonecrop growing at Rock Island Landing

A misty morning at Rock Island Landing

The quiet side channel between the west shore and Little Rock Island

A widow skimmer dragonfly along a Willamette alcove

A madrone in the Willamette Narrows

The view of Willamette Falls from the Highway 99E overlook

The downriver doors of Willamette Falls Locks

Looking downriver from the top of the fish ladder

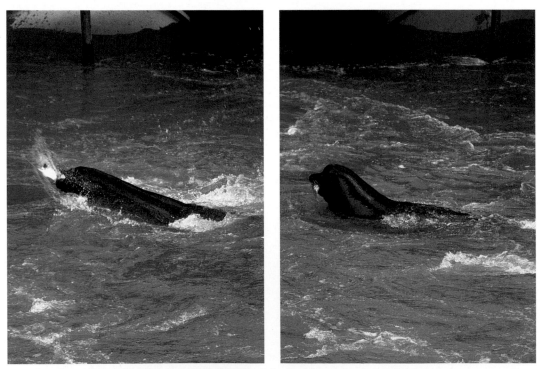

A sea lion slaps a fish against the water at Willamette Falls, then later takes its last few bites.

Vibrant and cool, the Clackamas River is the major tributary to the Lower Willamette

River meets parking lot at Cedar Island Boat Ramp.

An aerial view of Hogg Island

Amid the oak forest on Hogg Island

A mix of native and invasive plants grow on Hogg Island.

The small embayment on the main-channel side of Hogg Island

Mud-rich water after a downpour exits Tryon Creek on its way to the Willamette.

Ancient basalt on Elk Rock Island, with the Elk Rock cliff face across the river in the background

From Elk Rock Island's shore, you can look back toward its forest, which appears as an island of green on the basalt foundation.

The confluence of Johnson Creek and the Willamette

A restoration planting just up from the confluence on Johnson Creek

Snags along the river's west shore, with the Sellwood Bridge in the background

Stars race by overhead and airplanes cross the sky as the river moves northward under the Sellwood Bridge.

Oaks Bottom attracts a great many birds throughout the year.

Looking northward at the Ross Island complex, with East Island at the lower right. The berm connecting Ross and Hardtack Islands is in the foreground.

Autumn light bathes the old pilings in Holgate Channel. A small seep sends water flowing into the river from Oaks Bottom Wildlife Refuge.

Crows enjoy the beach near the large mudflat at the southern edge of Ross Island.

Old scrap metal can be seen along the shore of Ross Island.

The beach along Ross Island's west shore is composed of large rock and sand, mixed with a variety of detritus from the Willamette's high flows.

Ross Island makes a great destination for paddlers of all ages.

A riverside planting on the east bank in Portland

A barge unloads grain just downstream of the Steel Bridge. The Broadway and Fremont Bridges can be seen downriver.

Swan Island as seen from the heights of Forest Park

Viewed from Willamette Cove, a ship makes its way upriver.

Governor Ted Kulongoski (second from left) works with others to replant the McCormick and Baxter site in 2006.

Golden autumn light soaks the riparian fringe of Multnomah Channel.

Dusk falls on a November evening in Multnomah Channel.

Driftwood and sand at the beach at Kelley Point, with the Columbia River in the background

North American River Otter

River otters (*Lontra canadensis*) also live along the Willamette, although they are less often seen than beavers. Otters are semi-aquatic animals, with long bodies, webbed feet, rounded heads, small ears, and thick tails that taper at the end. Like beavers, they have thick insulating fur that is usually dark brown. Their diet is varied, but they are known to consume amphibians, fish, turtles, crayfish, crabs, and other invertebrates. They can also consume certain birds, their eggs, and some small mammals.

Along the Willamette, river otters are usually first spotted at a distance, appearing as something sleek and dark moving in the water. They often resurface just downstream of where they were spotted, with their dark eyes and whiskers peering above the surface. Otters tend to keep a good distance from people and remain in the near-shore areas, but there are certainly exceptions to the rule. On occasion, you might get lucky and see one tearing at a fish, then plunging down into the water only to resurface a few moments later to continue its meal.

Scientists have been studying river otters of the Pacific Northwest and have observed impacts of pollution on their endocrine systems. Chemicals, such as PCBs from legacy sources, can disrupt the reproductive capacity of male river otters. Given that toxic chemicals can still be found in the Willamette system, ongoing research is warranted to ensure that any hotspots are located and cleaned up.

River otters can be spotted along most of the Willamette's extent. Copyright N. Duplaix

5

Albany to Salem
RM 120 to 84

*Sure, we've made progress . . . but the old abuses
will come creeping back if we don't keep a careful,
constant watch over these waters.*

—MEL JACKSON

A STRETCH OF relatively calm water is found where the city of Albany spreads along the Willamette. Here the river curves eastward and then north again on its way toward Buena Vista. In the late nineteenth century, it was common to see steamboats traveling the river near Albany, transporting goods and people. Like many other Oregon towns, Albany was built on a foundation of logging. The river was also used as a key element of the natural resource industry, a home to large log booms and the related processing of the felled trees. While Albany has moved forward and developed other economic opportunities, today the Willamette is still used by the wood products industry, which takes a vast quantity of water from the river to process wood and returns it in the form of treated wastewater.

Albany's park system has been aggressive in obtaining new public land along the river. Former director Dave Clark sought opportunities to purchase properties from willing landowners, as part of a broad vision of what a small but growing town could do to serve its citizens while providing addition riparian habitat and undeveloped riverside areas.

The principal of public trust is that the public owns common or shared environments such as rivers, tidal areas, fisheries, and migratory species. The rights found in the public trust are said to be natural rights for all. Most rivers are publicly owned resources held in trust by public trustees. In the case of the Willamette, these trustees include the state of Oregon, the Oregon Department of Environmental Quality, the Oregon Department of Fish and Wildlife, and the Department of State Lands. There is a robust expectation that as public resources, rivers like the Willamette will be protected, especially concerning clean water and healthy habitat for fish and other wildlife.

The public trust doctrine has ancient origins. Roman law stated that the air, running water, and the sea and seashore were common to all people and should be protected. The Magna Carta of 1225 clarified the king as the protec-

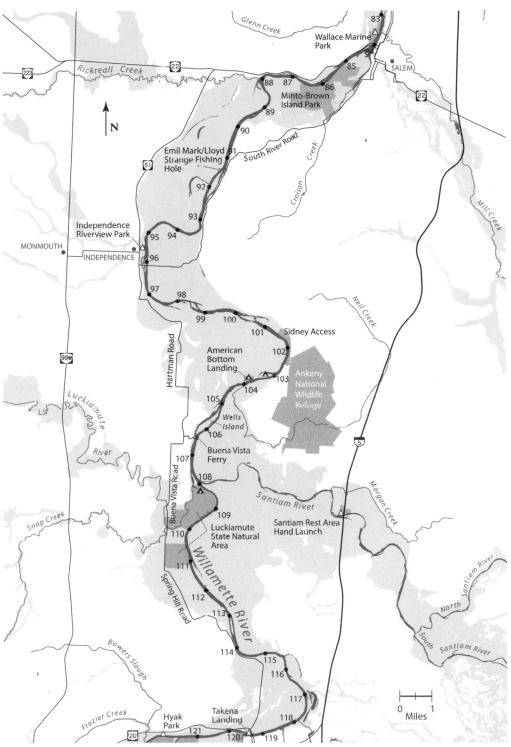

Albany to Salem

Public riverside lands like this Department of State Lands property are key resources along the Willamette.

tor of public and common rights in Britain. The king could not sell public trust resources to private interests. The excellent book *The Riverkeepers* by John Cronin and Robert F. Kennedy Jr. provides an excellent in-depth discussion of public trust.

Among other laws, the Clean Water Act and the Endangered Species Act represent the federal government's role in protecting clean water, plants and animals, and their habitats. The regulatory force underlying the Clean Water Act is related to the permits that are issued to businesses and municipal wastewater treatment facilities. Ensuring that their discharges are as clean as possible is a critical part of upholding the public trust. If rivers are sullied by contamination and species contain pollutants that affect their health and the health of people, then the public trust is not being upheld. Over the years, the doctrine has been disregarded in many cases along the Willamette, and laws like the Clean Water Act have been poked with so many holes and bent by creative interpretations that the public trust is not being protected to the full extent necessary.

Although there is a legal basis for protecting the Willamette, there is much yet to be done with regard to the cleanliness of the river water and bottom in some areas and healthy habitat in the entire system. Some people seem to believe that the Willamette exists simply to be used and manipulated in any manner that serves their interests. Although rivers can and do provide many valuable resources, we now understand that clear lines need to be drawn to protect the public trust. Over the years, many bad choices have been made that plague our nation's river systems, and these can provide valuable lessons to inform our actions today. The dredging of rivers is one such action.

Since the 1990s, a vocal minority has sought to dredge the Willamette River. The group behind this effort has only one thing in mind, to enable large powerboats and riverboats to travel from Newberg to Corvallis and beyond.

While riverboat travel conjures romantic notions of the Willamette and its relatively brief steamboat era, this type of craft probably does not need to travel the most dynamic stretches of the river today. Several decades ago, most people would not have thought twice about dredging the river to make the mainstem just a little bit deeper. Perhaps just carving off the top of some of those riffle-creating gravel bars that hinder upstream movement of large boats in the summer would make life easier. More recently, however, hydrologists, biologists, and other river experts have widely acknowledged that dredging rivers to meet travel needs or to control river flows has too often damaged the health of river systems. Dredging affects fish habitat, releases sediment downstream, and can alter the flow of cool water to the river. Large rivers like the Willamette carve and build gravel bars naturally, an important part of the natural river ecosystem that helps sustain aquatic life. The Willamette bears the marks of many such actions. Dredging is also a poor economic move, given that the river will build additional gravel bars over time that will require further dredging. In addition, there are often unintended consequences of changing the natural flow patterns of rivers. One need only look at the legacy of hydropower dams to realize this fact.

Another question to ponder is what is appropriate in a river like the Willamette. The large reservoirs behind dams already provide ample places for powerboats for waterskiing and wake boarding. From the river's confluence with the Columbia, powerboats with propellers can travel the deep waters 60 miles upriver. A strong argument can be made that there is plenty of deep water for powerboats, without the need for the channels of the Mid and Upper Willamette to be dredged. These dynamic shallow stretches of the Willamette are important habitat for fish and other wildlife, and they can be navigated by any person with a little expertise and a paddle craft. Also, it can be argued that dredging the river between Newberg to Corvallis would break the public trust doctrine.

Just above the Willamette's confluence with the Santiam River at RM 108, a riffle has been sculpted by the push of the Willamette's spring and winter flows, creating some small jumping waves at low summer flows, with a relatively narrow band of fast-moving water on river left against the bank. To the right the waves slowly dissipate and a slow eddy emerges. The eddy moves upriver toward a gravel bar, where the water assimilates with the filtered current moving downstream again. Under the proposals to dredge the Willamette, such areas would be gutted, with hundreds of tons of gravel scraped from the river bottom, all for the sake of a slow-moving tourist boat from Salem and a few more miles open to wake boards. Dredging costs time and money that would be much better spent improving the healthy ecological function of riverside areas that support native wildlife. If the public's ability to see and experience the Willamette is in question—and some cannot see the river by drift boat, canoe, or kayak—then motorized jet boats might provide the solution. Many of these craft are designed and built in Oregon, and jet boats are regularly used by fishermen to travel the river.

Luckiamute State Natural Area, RM 112 to 108

WHERE: From Independence, take Corvallis Road south to Buena Vista Road and head south. From Albany, take Springhill Drive to Buena Vista Road and head north. There are three access points from Buena Vista Road. One is north of the Luckiamute River, on a property owned by the Oregon Department of Fish and Wildlife, which provides a footpath toward the winding Luckiamute River. The next access is immediately south of the Luckiamute River; a short drive down a gravel road takes you to a trailhead and parking area. The last road access is nearly to River Road, with a small parking lot and additional trail access.

AMENITIES: Parking areas and trails.

With a total of 918 acres, the Luckiamute State Natural Area is a large tract of lowland forest, wetland, and oak savannah that provides an abundance of habitat for a range of native species. Bring binoculars for the good birding opportunities at this site.

Luckiamute State Natural Area is situated above the Willamette's confluence with the Santiam and the smaller Luckiamute River just downstream. These two rivers are quite different from each other. The Santiam River joins the Willamette on the east side, and the river's North Fork and South Fork each have their source waters in the high elevations of the Cascades. The river pours in a great volume of relatively cold and clear water from the 1818 square miles of its basin. The headwaters of the Luckiamute are in the Coast Range, and it is typical of this river to have low flow volumes from June through October. The total area of the Luckiamute River Basin is 368 square miles, and the river is 108 miles long. Each of these rivers provides different habitat for wildlife, yet together they provide a relatively complex ecological stronghold. With two confluence areas so close to one another, in addition to the main river, the area provides an abundance of habitat for a range of fish species, as well as birds and mammals that are linked to them. Water temperature varies from colder where the Santiam flows in to warmer where the Luckiamute enters the stream.

At the Santiam's confluence, a significant amount of water can be seen mixing with the Willamette; depending on the season, swirling eddy lines and upwellings are common downstream for up to a half mile. The Luckiamute River's flow is less noticeable most of the year. At first glance, the Luckiamute appears as a simple backwater, rich with vegetation and visual beauty. Yet when the river is high, you can easily see its flow mixing with the Willamette, with its own collections of strong eddy lines and upwellings. The varying currents of the two rivers intermingle with gravel bars and woody debris to produce a dynamic river bottom environment and an abundance of good habitat for spring chinook as well as nonnative species.

The natural area begins 2 miles upriver of the Santiam River confluence. The southern section of this two-part natural area contains a large expanse of western pond turtle habitat that is being improved by restoration work, in cooperation between the Oregon Parks and Recreation Department, which owns the property, and the Oregon Department of Fish and Wildlife. There is limited public access to the area via trails, and the primary goal of this section is to help restore the turtle's population size. This section of the Luckiamute State Natural Area also has wetland and pond habitat suitable for birds such as great egret, which can be seen fairly often, as well as green heron and American bittern (*Botaurus lentiginosus*), a medium-sized, stocky heron with brown, tan, and white stripes.

A private agricultural property separates the southern and northern tracts. If you paddle a canoe or kayak to the landing just upstream of the Santiam confluence before the riffle, you can gain access to a trail that threads its way through the cottonwoods. After about a mile, the trail leads into an open area that was once cultivated. Most of the northern tract is the original Luckiamute Landing greenway park. This section is comprised of some open land that was used for agriculture, but the majority is lowland forest full of black cottonwood, Oregon ash (*Fraxinus latifolia*), Pacific willow, bigleaf maple (*Acer macrophyllum*), black hawthorn (*Crataegus douglasii*), and others. The shrubby red osier dogwood (*Cornus sericea*) and other native plants such as snowberry (*Symphoricarpos albus*), trailing blackberry (*Rubus ursinus*), camas (*Camassia quamash*), and Dewey sedge (*Carex deweyana*) can be found here. In lower areas of the park one can see Pacific ninebark (*Physocarpus capitatus*), Nootka rose (*Rosa nutkana*), salmonberry (*Rubus spectabilis*), vine maple (*Acer circinatum*), red elderberry (*Sambucus racemosa*), oceanspray (*Holodiscus discolor*), and the notorious Pacific poison oak (*Toxicodendron diversilobum*).

This portion of the Luckiamute State Natural Area gives you a real sense of what some of the original floodplain forest along the river might have looked like. From the air the northern area is pure trees, with only the mainstem Willamette stopping the forest's eastward advance. To the west the natural area recedes to the farmland that forms a clear boundary and extends nearly as far as the eye can see outside the park boundary. Much of the Willamette Valley, with the notable but relatively small exceptions of parks such as Luckiamute State Natural Area, is now farmland. A good portion of this geometric agricultural landscape was formerly oak savannah, with much more lowland forest near the river. ■

When the Oregon Parks and Recreation Department was developing the master plan for the Luckiamute State Natural Area, a handful of local landowners felt that their rights were being deprived. A new public access point to the Luckiamute River was proposed, as well as significant restoration efforts that might eliminate agricultural activity within the park boundary. As with several other state parks, some land is leased to local agricultural producers. Such a scenario provides a level of fairness to farmers and can help manage lands until restoration activity can begin. In most cases, the lease of the land for agricultural production is for a limited time, with priorities for public property turning toward the improvement of native habitat for fish and other wildlife. The vast majority of people seem to understand that properties like Luckiamute State Natural Area are rare gems amid a sea of towns, cities, and agricultural land. Efforts to protect and improve their condition for native species and to allow public access are integral aspects of publicly owned properties along the Willamette.

In higher areas of the Luckiamute State Natural Area, Oregon white oak savannahs can be found. This upland habitat type has been greatly reduced throughout the Willamette Valley. Historically oak savannah was quite common, but today there are only remnants left. Imagine large prairies of grass, dotted with massive oak trees expanding from the foothills to near the river bottoms. Lands with small patches of such habitat are prized today, giving increased importance to protecting the Luckiamute State Natural Area and restoring areas that show the promise of becoming viable oak savannah habitat.

Bird life in the spring is rich at the natural area, which serves as a much needed area for at-risk bird species. Willow flycatcher (*Empidonax traillii*), yellow warbler (*Dendroica petechia*), purple martin (*Progne subis*), Oregon vesper sparrow (*Pooecetes gramineus* ssp. *affinis*), horned lark (*Eremophila alpestris*), common nighthawk (*Chordeiles minor*), western meadowlark (*Sturnella neglecta*), chipping sparrow (*Spizella passerina*), and others find homes in this natural area or nearby. Luckiamute State Natural Area, with its trail access points, can provide birders a great opportunity to view a range of species.

The Luckiamute State Natural Area is part of a renewed strategy by the Oregon Parks and Recreation Department to protect its existing holdings along the Willamette River. The Willamette River Greenway Parklands Strategy codifies the previous goals of the Willamette Greenway Program and sets forth the department's strategy to continue and bolster this program on behalf of the public. Much of the land along the Willamette has been transformed from its original lowland forests and grasslands. As reflected in the Luckiamute State Natural Area, there is promise for restoration work along the river to provide tangible benefits for wildlife and, in turn, for people as well.

While the Willamette Valley is home to a good share of bird species, there are those, like the yellow-billed cuckoo (*Coccyzus americanus*), that no longer reside here. According to *Birds of Oregon* (Marshall et al. 2003), until the 1920s the yellow-billed cuckoo was once abundant in the willow bottoms along the Willamette and Lower Columbia River, but this vibrant population steadily declined. Today only sporadic sightings of the yellow-billed cuckoo are reported, typically in eastern Oregon. Ornithologists believe that the near disappearance of this species in its former range can be attributed to the extensive loss of riparian habitat. The yellow-billed cuckoo has a large range requirement, meaning that it takes a large area with significant stands of riparian forest to sustain a breeding pair. The western subspecies is a candidate for listing under the federal Endangered Species Act. The Willamette Valley has been discussed as a possible region for restoring breeding populations of the yellow-billed cuckoo. Once the Luckiamute State Natural Area is further improved, with more lands acquired and protected within a half mile of the river, it would serve as a fine place to reintroduce this species.

At-Risk Fish in the River

The Willamette River has thirty-one native and twenty-nine nonnative species of fish. According to the Pacific Northwest Ecosystem Research Consortium, several of the native fish are listed by federal or state agencies as endangered, threatened, or sensitive: spring chinook (*Oncorhynchus tshawytscha*), steelhead trout (*Oncorhynchus mykiss*), Oregon chub (*Oregonichthys crameri*), coho salmon (*Oncorhynchus kisutch*), cutthroat trout (*Oncorhynchus clarki*), and bull trout (*Salvelinus confluentus*). Spring chinook are present in the Willamette from both naturally producing and hatchery stock. Some of the main fish sought by fishermen along the river include spring chinook, cutthroat trout, white sturgeon, steelhead or rainbow trout, smallmouth bass, and crappie. The latter two are invasive exotics.

One of the state's endangered fish, the Oregon chub, is a minnow. It has an olive-colored back grading to silver on the sides and white on the belly. Adults are usually less than 3.5 inches long, and they feed on the tiny larvae of mosquitoes and other insects. Oregon chub used to be distributed throughout the Willamette Valley in off-channel habitats with little or no water flow, silty and organic substrate, and abundant aquatic vegetation and cover for hiding and spawning. Historically, the mainstem of the Willamette was a braided channel with a lot of suitable habitat for the chub, including beaver ponds, oxbows, stable backwater sloughs, and flooded marshes. In the last century, these habitats have disappeared or degraded due to the construction of dams and revetments, channelization, drainage of wetlands, and agricultural practices. This loss of habitat and the introduction of nonnative predators, such as bullfrogs, largemouth and smallmouth bass, and crappie, have resulted in a sharp decline in Oregon chub populations.

In 1993 this fish was listed as endangered under the federal Endangered Species Act. Currently, populations of Oregon chub live in the Santiam River, Marys River, Long Tom River, Muddy Creek, McKenzie River, Coast Fork, and Middle Fork drainages, with most in the Middle Fork. In 1998 the U.S. Fish and Wildlife Service published a recovery plan for the Oregon chub, and the U.S. Forest Service, U.S. Army Corps of Engineers, and Oregon Department of Fish and Wildlife have active programs to protect this minnow. The goal is to reverse the species decline by protecting existing wild populations, reintroducing chub into suitable habitats throughout its historic range, and increasing public awareness and involvement.

A mile below Luckiamute State Natural Area is the small landing of Buena Vista. The landing itself dates back well over a hundred years to a time when steamboats once plied the Willamette, with their shallow bottomed hulls and steam-powered paddle wheels pushing the craft up the Willamette's swift current when flows were high enough. The Buena Vista Ferry is located at RM 106.5. This ferry is typically open only a few days a week (check with Marion County for times). It was once typical for riverboats to pull up to small towns along the Willamette and take on people and agricultural goods. In Wheatland and other early riverside towns upstream and down, the riverboats were

At Buena Vista, a small ferry travels back and forth across the river many times daily.

counted on as a critical link to the rest of the world. The ferry in operation at present, which travels the short distance from one side of the river to the other, enables local residents to save time in getting across the river.

Such cross-river ferries have difficulty operating at higher river flows, because the river exerts an amazing level of force and the ferry is essentially held in place by two cables. A ferry of this sort has a relatively benign impact on the river's health. As one of only three ferries operating on the Willamette River, the Buena Vista Ferry may not last much longer, given the stiff competition for funding and changing priorities in both Marion and Polk Counties. One alternative that has been suggested is to build another bridge over the Willamette, somewhere in the general area. Relatively few people, however, would be served by an additional bridge between the cities of Independence and Albany.

Downstream of Buena Vista the river splits, with the main channel heading to the right and a secondary channel flowing left around Wells Island at RM 106. This island and associated backchannel is very picturesque. The island, owned by Polk County, once had camping facilities, but these were washed out in the flood of 1996. The flow in the side channel can be a bit startling for paddlers, as unpredictable eddy lines and upwellings move about. The entrance requires particular attention, as strong eddies are found at some flow levels, and the channel is punctuated with woody debris such as large logs and root wads. Further down the channel is a cliff face to the left, with the coursing river moving left, then right, and left again. While this short backchannel has creative flows, its scenery is worth the extra time to get around the island.

High banks like this one just up-river of Buena Vista are few and far between along the Willamette.

Albany to Independence, RM 120 to 96

> **STARTING POINT**: Put in at Bryant Park on the south side of the river or at Takena Landing; both sites are in downtown Albany.
> **ENDING POINT**: Independence Riverview Park, located on the river in downtown Independence.
> **DISTANCE**: 24 miles
> **SKILL LEVEL**: Intermediate skill, with experience on fast-moving water.
> **CONDITIONS AND EQUIPMENT**: This is a large open section of river with only a few areas of fast moving water.
> **AMENITIES**: There is parking and restrooms at both Bryant Park and Takena Landing in Albany. By calling ahead, you can get permission to park overnight. Independence has free parking and restrooms.
> **WHY THIS TRIP?** You can see an abundance of wildlife and experience the confluence of the Luckiamute and Santiam Rivers near Luckiamute State Natural Area. This trip can be done overnight, as there are campsites along the way.

The first few miles of this stretch are calm, as the river gently pushes east then northward. After 9 miles of paddling you will see a large stand of cottonwood situated at Luckiamute Landing. A point of interest is the Santiam and Luckiamute Rivers confluence within the Luckiamute State Natural Area.

Below Buena Vista the river is wide and scenic. American Bottom Landing at RM 104 is a campsite owned by Oregon Parks and Recreation Department. This low-lying area can make an excellent place to stop, although it bears the full force of the river during high flows and can be littered with woody debris at other times. This is another area that seems almost from another time, given its quiet setting and abundant riparian forest.

Near RM 100, the hills just south of Salem can be seen. As Independence draws near, homes can be seen along the river, first on the east side, then on the west. The Independence Bridge indicates the town just downriver. Independence Riverview Park makes a good put-in and take-out, with the downstream end of the park working well for paddlers. The boat ramp can be very shallow. In this case it makes sense to remove gravel, so that those who wish to navigate the Willamette can do so without damaging the riverside. ▪

Early in Independence's history, it was served regularly by steamboats. Today, a stone's throw from the main river, several restaurants and shops maintain the historic downtown core. Recently, the city has begun to turn back toward the river by making improvements to Independence Riverview Park, including restrooms and an amphitheater. In time the nearby gravel operation is slated to move, which could provide an opportunity to secure more

Bald Eagle

A juvenile bald eagle looks down at the river.

The bald eagle (*Haliaeetus leucocephalus*) is a very large raptor whose long broad wings are held flat while soaring. Bald eagles range in length from 28 to 38 inches and have an amazing 80-inch wingspan. Adults are easily recognized by their brown body and white head and tail. Juveniles, however, are patterned with various amounts of dark brown and white throughout. It takes 5 years for a bald eagle to acquire its full adult plumage. Juveniles might be confused with golden eagles (*Aquila chrysaetos*), which have similar all-brown plumage patterns, but golden eagles have feathered legs and they lift the outer parts of their wings when soaring.

Bald eagles eat large birds and mammals, but they prefer fish. They also scavenge on dead carcasses. They occasionally hunt cooperatively, with one eagle flushing prey toward another, and they wade in water to catch fish. Bald eagles also attempt and sometimes succeed in stealing prey from other predators, such as ospreys.

Bald eagles can be seen throughout the Willamette River system. In 2007 the bird was taken off the Endangered Species List, where it had been listed as threatened. Bald eagles are not numerous, but the watchful river traveler can see them, whether traveling near Eugene, Salem, or Portland. Their nests are usually built in the top third of cottonwood trees, with a large collection of sticks typically placed where the trunk diverges into a Y shape. While you may get a good view of a bird perched high in a tree next to the river, always take care not to get too close to an eagle's nest, as the parents may fiercely defend their eggs or hatchlings.

public land along the river. Downstream of Independence is another good example of a Willamette River floodplain. For several miles, the land to the west of the river is flat and open. Today this floodplain is used for the production of grass seed and other agricultural crops.

The river winds its way slowly past a large island at RM 92, which is owned by the Department of State Lands, and a somewhat popular Polk County fishing hole. AT RM 89.5 is a small publicly owned island that can be used as a stopover, the last quiet spot for a few miles given the proximity to Salem. Throughout this stretch of the Willamette traffic can be heard, on river right from River Road and eventually where the river turns sharply to the east and

Common Merganser

The common merganser (*Mergus merganser*) can be found all along the Willamette from Eugene to Portland, although it is not a common bird. This colorful diving duck is about the size of a mallard. The male has white sides, a black back, and an iridescent green neck and head. The female has a light gray body, a rust-colored head, and a white chin patch. Both sexes have bright orange beaks.

A male and female common merganser on the Willamette

Typically you can see common mergansers next to the riparian fringe, where in the summer months a female and her young may be seen swimming away from you. If you approach, they will typically fly off in a hurry only to land a bit downstream. If you continue to travel toward them, mergansers often simply repeat their hurried flight. At times you can also find common mergansers sitting on logs and other riverside structures. Mergansers eat a range of fish found in the Willamette, from sculpin and suckers to lamprey, usually juveniles that are less than 8 inches long. As a result, you will typically see mergansers in areas where the water is deep enough for them to dive. Their nests are typically in tree cavities, brush, rocky ledges, hollow logs, and nest boxes.

Sticks stripped of their bark by beavers near Minto-Brown Island

parallel to Highway 22. Running toward Salem, the river passes a gravel quarry and several riverside homes. These final miles into the state capitol maintain the dual nature of the Willamette: people zipping along on the noisy highway as the river quietly flows along.

Starting at RM 87, the large open landscape of Minto-Brown Island Park can be seen. This park is owned by the city of Salem and is 898 acres of floodplain and open woodland area. The city acquired the initial Minto Island property with funds from a Land and Water Grant and the Willamette Greenway Program. The landowners also made a donation so the purchase could occur. A year later, Marion County acquired a 525-acre site next to Minto

Island. Some of the land was originally owned by Isaac "Whiskey" Brown, who made his home on one of the islands. John Minto purchased the adjacent island, which was named after him. These were classic Willamette floodplain islands covered with vegetation and later cleared for agricultural use. The Willamette's course once clearly divided the islands, with Minto's island on the east bank and Brown's island on the west bank. The flood of 1861 changed this pattern, for when the flood water subsided, the river had bounced to the north and west.

Minto-Brown Island Park provides an excellent area for riverside recreation such as birding. A good portion of the island has undergone significant tree planting. Although this park is right on the edge of the city, it can provide valuable habitat for wildlife. A very nice backwater makes its way around part of the park from RM 84.25, with the channel extending about a mile inland. Here it is common to see a variety of riverside birds and even an occasional egret. This park provides a relatively quiet and scenic area close to the hustle and bustle of the state capitol.

Wallace Marine Park, on the west side at RM 84, provides a good put-in and take-out point for access to Minto-Brown Island and for longer excursions as well. At higher flows, it works well to use the gravel bar and backwater area, which is located just downriver of the highway bridge. Nearby you can see the former railroad bridge. Recent discussion has centered on making this a pedestrian river crossing.

The large gravel bar at the old boat ramp at Wallace Marine Park makes a fine put-in for canoes and kayaks or a place to just stroll downriver.

6
Salem to Newberg
RM 84 to 50

Do we have the right to ask why more
has not been done by more people?
—Tom McCall

SALEM and the Oregon state government have always been connected to the health of the Willamette. The state capital, home to such state agencies as the Department of Fish and Wildlife and Oregon Watershed Enhancement Board, has played an active role in the Willamette's issues and improvement. As recently as the mid-1990s Governor John Kitzhaber and his administration worked hard to increase habitat restoration and improve water quality for the Willamette and other rivers through the creation of the Oregon Plan for Salmon and Watersheds, and helped to fund significant restoration work along the Willamette and its tributaries. Kitzhaber also created the Willamette Restoration Initiative which brought together a variety of stakeholders to identify key issues and actions that should be taken in the Willamette system.

In 2002 when Governor Ted Kulongoski was voted into office, he continued to focus on the Willamette with his Willamette Legacy Program. This effort by Kulongoski and his staff helped to make additional progress for the river, fed by increased state investment in habitat restoration, water-quality monitoring, and recreational improvements such as the Willamette Water Trail.

Thus after a relatively long period of inaction, Salem's connection to the Willamette is now long-standing, with each administration renewing efforts to improve both water quality and habitat. The state of Oregon has been instrumental in creating change, especially on the habitat front, though there is also a renewed focus on toxics in the river system. In 2007 the Oregon Watershed Enhancement Board dedicated several million dollars of its Oregon Lottery funding to river restoration. In tandem with this effort, the Meyer Memorial Trust created a program that also dedicated millions to habitat restoration on the Willamette. The hope is that these state and private funds will leverage additional federal funds for restoration and water quality improvement.

Water quality defines the relative cleanliness and purity of water. Water quality measures are most commonly applied to a river system to help categorize its overall health and its impact on wildlife and people who come into

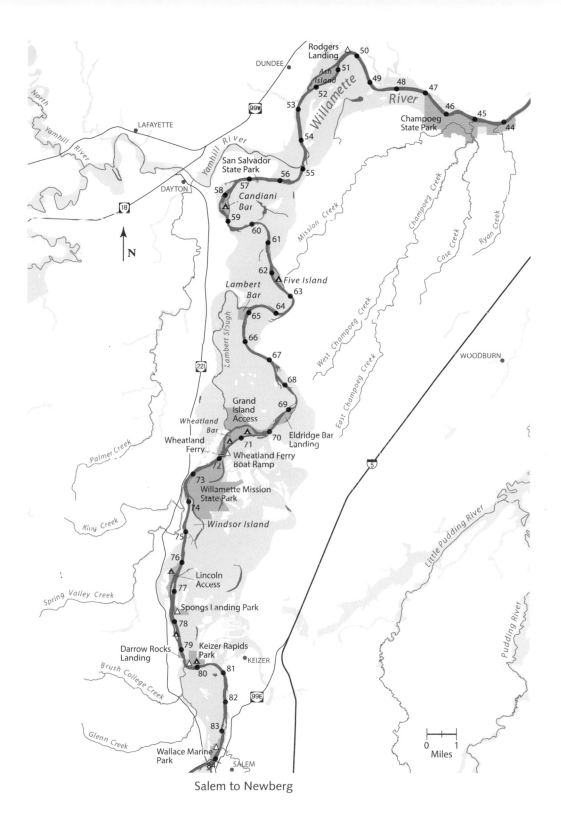

Salem to Newberg

contact with the river. In the 1960s, water quality issues were easy to see. Fish dying in the river were a vivid image that the average person could easily understand. Over the years, most water-quality issues were related to the cities that line the river from Eugene to Portland, typically associated with municipal waste facilities and major industries such as the wood pulp and paper industry. These sources have historically provided the most significant volume of polluted water to the Willamette. Because this type of pollution comes from specific pipes, or point sources, they are relatively easy to regulate as agencies can measure the cleanliness of the treated wastewater. In contrast, nonpoint sources include streets, fields, and other areas where water can flow across the land to a creek or river, and these are not easy to monitor. Consequently, early pollution reduction efforts in the Willamette Valley focused on point sources and ensuring that municipal treatment plants had the capacity to provide both primary and secondary treatment.

Improving treatment of point source pollution was the big leap forward for the Willamette, as chronicled in the June 1972 *National Geographic*. Led by Governor Tom McCall and building on the work and leadership of countless others over the years, Oregon's government made a conscious decision to improve water quality in the river. McCall can perhaps best be characterized as the right voice at the right time, given the decades-long interest in dealing with the river's water quality. His unique passion and perspective helped keep the Willamette's cleanliness in the public mind, thereby building political support.

Many times the river's cleanliness was brought to the fore by local people, typically angered by its poor condition. In Portland in the 1920s, local officials closed the river to swimming on numerous occasions, and subsequent reports by the City Club of Portland found the river to be filthy, with the city's own sewage as a major source of the problem. Fish kills had become numerous as well. In 1938 the political momentum was generated to create the Oregon Sanitary Authority, as a result of a ballot initiative and the work of civic leaders such as Portland Mayor Joe Carson. Unfortunately, creation of the Oregon Sanitary Authority did not do much to improve conditions along the Willamette throughout the 1940s. After World War II, however, the Oregon Sanitary Authority required better waste treatment facilities to be built in Eugene, Harrisburg, Salem, and Portland. With these improvements in place, it became apparent that discharges from wood pulp and paper mills were still a problem.

Much of this progress in the 1930s and 1940s can be attributed to the work of Dr. David Charlton, a lead figure who fought pollution as part of the Izaak Walton League and operator of Charlton Laboratories. In 1948 the league published a report that indicated too little had been done to address the impact of pulp and paper plants. For many years, Charlton was an ardent advocate for the river's health, seeking to inform people of the level of pollution along the river and pushing for more to be done.

The Oregon Sanitary Authority engaged the pulp mills and in 1950 set a deadline for them to stop dumping wastes into the river, specifically sulfite liquors, as of May 1, 1952. Although the mills sought other uses for these by-products as an alternative to dumping wastes into the river, change came slowly. The Oregon Sanitary Authority extended deadlines as the companies made little progress. The mill owners argued that the cost of waste treatment would cause a significant loss of jobs and higher prices for wood and paper products. Such arguments would be repeated for decades as companies were directed to do more to improve their wastewater treatment. In many cases, they fought additional improvements with propaganda. The pollution problems continued into the 1960s.

In November 1962 "Pollution in Paradise," a documentary created by journalist Tom McCall, was aired on KGW Television. McCall had been interested in the river's issues for some time and had delivered commentaries that described some of the Willamette's pollution issues. McCall proposed that Oregonians had an "environmental malaise" in relation to the river. He planned to do a more in-depth commentary on the Willamette, and when he jumped into his research for the project the resulting information was shocking. Part of McCall's surprise was related to what the pulp and paper companies were allowed to do. He and his crew got footage of direct discharges of pollutants by numerous mills along the Willamette and investigated the impact these had on the river.

In "Pollution in Paradise," McCall showed that the Willamette River was actually cleaner when the Oregon Sanitary Authority was created in 1938 than it was in 1962. In addition to industrial pollution from the pulp and paper industries, the documentary focused on sewage dumped from sources ranging from municipal waste systems to houseboats. McCall spoke to an industrial representative who stated that river pollution was only a "small inconvenience" to sustain a healthy economy. Therefore, another significant aspect of McCall's report was to make the case that Oregonians did not have to choose between jobs and a healthy river, that livability was more important in the long run to the state's economic viability.

McCall's report incensed the general public, putting the river's wretched condition on clear display for thousands of Oregonians. As a result of this work, wheels were put into motion that would propel Tom McCall's political career and in tandem make advances for a cleaner and healthier Willamette River. Soon after the report was aired, a state law was passed to provide the Oregon Sanitary Authority with the power to shut down polluting facilities.

Unfortunately, it would be several years before the Oregon Sanitary Authority would use this power. For years, the U.S. Army Corp of Engineers provided enough flow during the summer months to lessen the pollution problems. The dry weather of 1965 resulted in much less flow, however, making the river's ongoing pollution problems apparent.

Tom McCall became the secretary of state in 1964 and was elected governor in 1966. Once in office McCall went into action on the Willamette, demanding results from the Oregon Sanitary Authority. In 1967 the sanitary authority required secondary treatment at all municipal facilities and enacted new measures for industrial facilities. McCall soon took a bold step and appointed himself as member of the Oregon Sanitary Authority, eventually becoming its chair and giving it some much-needed backbone.

The Department of Environmental Quality was created from the Oregon Sanitary Authority in 1969, and McCall named L. B. Day its first director. While most industrial facilities went along with requests to improve their discharges, the department clearly needed the legal authority to punish those who polluted the river. More stringent permit requirements and the provision for the department to revoke permits and even close plants brought about a new atmosphere. After this effort, the Willamette River's bacteria and dissolved oxygen issues improved dramatically. There are plans to erect a statue of Governor Tom McCall in Salem's Riverfront Park, a fitting tribute to the visionary leader who not only helped to improve the Willamette's overall condition, but who also made great strides for public beaches and land use planning as well.

These advances made in the water quality of the Willamette were significant, in some ways setting a national standard for what could be done to turn around a river's condition in the period of a decade. As important as the great clean-up during McCall's tenure was, we have learned much in the past 35 years about the overall cleanliness of water and the surrounding landscape along the river. Today, water quality is measured at various levels. Basic measures, such as water temperature, dissolved oxygen level, conductivity, and pH can indicate much about the quality of a water sample. In addition to these measures, however, sampling for bacteria is very important to understanding the health of the river's water. Because sources range from wildlife feces where large flocks of geese congregate to sewage spills and related releases, many areas keep tabs on the level of *Escherichia coli*, an indicator of other bacteria. The concern with *E. coli* levels is that people can ingest water containing high levels of bacteria while recreating along the Willamette, which can lead to serious intestinal illness. This is the principal human health issue in relation to water quality along the Willamette.

With the breadth of activity in the Willamette Basin, from its urban centers to agricultural lands, more people have become interested in toxic pollution. Since the early 1970s, significant strides have been made in understanding the interface between toxic chemicals, such as mercury and PCBs, and the river's water and sediment. Currently researchers are examining the effects of pesticides, the sources of mercury, how these chemicals interact with river wildlife, and how long they persist. Newly recognized pollutants are also being examined, including caffeine and pharmaceuticals that are not necessarily filtered out at municipal treatment plants. In general, the quality of the Willamette's water degrades from upriver to downriver. For instance in Eugene, the basic water quality indicators are in fair shape. When the river reaches Salem these indicators show declines in water quality, from higher temperature to less clarity of the water (turbidity). Continued sampling and testing of the Willamette's waters is still necessary on a regular basis.

In the mid-1990s, as part of the National Water Quality Assessment Program, the U.S. Geological Survey did some intensive sampling to test for toxic compounds in the river. The report indicated that pollution continued to threaten native fish, agricultural fertilizers and pesticides in streams and groundwater were degrading water quality, organochlorine pesticides and PCBs were detected in sediments and wildlife, and dioxins and furans were found in all sediment and fish tissue samples.

Dioxins and furans are organic chemicals that can be very toxic to wildlife. These compounds disrupt the endocrine systems of animals, which affects their ability to reproduce successfully and has serious impacts on the health of wildlife populations, with river otters being a good example. In certain areas the U.S. Geological Survey also found high concentrations of DDE, a by-product of DDT, a pesticide that was used widely from the 1940s until 1972, when its manufacture was banned.

In all, the report provided a sobering look at pollution in the Willamette system. This work has helped to guide efforts to curb runoff from farms, cities, and industrial areas and the continued scrutiny of wastewater permits. The report took special care to point out the difference between a stream in the headwaters of the Willamette system with good water quality, characterized by a healthy riparian area and shade, with the Little Pudding River, with absolutely no riparian cover in many areas.

Testing of the Willamette's water and sediment also occurs on a more piecemeal basis, as municipalities analyze the river's water quality related to drinking water or in specific areas where existing concerns have prompted action, such as Portland Harbor, where high levels of contamination exist, and Newberg Pool, where fish deformities occur. In 2004 the U.S. Geological Survey conducted a survey of the Clackamas River Basin that focused on household chemicals, and their results showed that pesticides and other chemicals were pervasive in that system as well. After many years with relatively few active projects to sample and test the Willamette's water and bottom sediments, in 2007 the state legislature provided funding for a toxic pollutant testing program to be administered by the Department of Environmental Quality. As the state agency in place to protect the public and wildlife against environmental pollution, the department will now have a program to sample and test water quality and sediment in an ongoing and carefully planned way. This program should expand our understanding of pollution sources and new trends that can harm the river as well as the wildlife and people who use it. In addition, it will help to ensure that policy decisions are based on sound data, which unfortunately has not always been the case.

As you move downriver from Salem, homes line the banks for the first few miles. This changes at RM 80, where you reach the 140-acre Keizer Rapids Park that provides some good riverside habitat for a range of species. Formerly an Oregon Parks and Recreation Department property, in 2002 Keizer Rapids was leased by the city of Keizer. Before dams controlled the Willamette's flow, during the dry months the river would routinely drop to levels that created some minor rapids at this point. At modern river levels, the river's main current moves over some shallow gravel and larger rocks on river left, creating a strong eddy line or two. This was also a point where, it is said, in historic times one could walk across the river, hopping along gravel bars and wading in the swift flow. Today the U.S. Army Corps keeps the flow in the Salem-Keizer area near 7000 cubic feet per second during the summer months, largely to dilute the treated discharges from municipal treatment plants.

Keizer Rapids Park was created by a significant community effort. For years, Keizer had only three small properties along the river, all with rather poor access for the average person. Working with the Oregon Parks and Recreation Department, the city obtained a long-term lease for the property. In turn, Keizer has developed a management plan for the park that will provide improved public access while keeping the site's natural attributes intact. Co-operative partnerships like this can help preserve and manage riverside lands, but perhaps more importantly they represent an increased level of interest that cities like Keizer have in preserving the Willamette's natural heritage. The Keizer City Council and Councilor Richard Walsh, who understand the importance of caring for the river, helped push this project forward for the benefit of Keizer and the greater Salem area.

Keizer Rapids Park, RM 80

WHERE: From Interstate 5 take Exit 260A, head west on Lockhaven Drive for about 1 mile. Take a left on River Road (Highway 219), and travel about a half mile south to Chemawa Road. Take a right and follow Chemawa Road into the park area. You will see a sign and a parking area to the left.
AMENITIES: Ample parking and portable toilets.

Keizer Rapids is a large new park that is in the process of being improved with trails and access to its natural area. From the parking area you can take a loop trail that makes its way through the property, providing access and views of the large gravel bar floodplain area, as well as the upland forest and wetland area. A nice transition from upland fir forest to the large gravel bar starts with black cottonwood, followed by Pacific willow. While this area provides for multiple uses, such as biking and dog walking, with an amphitheater planned, it also gives a good access point to the river and an opportunity to learn about its wildlife and habitat. ■

Wallace Marine Park to Willamette Mission State Park, RM 84 to 72

STARTING POINT: Wallace Marine Park in Salem, just west of the Highway 22 bridge. Use the gravel bar, rather than the boat ramp, as a put-in, which enables you to gather your gear and get prepared in a more relaxed setting.
ENDING POINT: From Interstate 5, take the Keizer exit (260A) and follow Lockhaven Drive to River Road (Highway 219). Head north for about a half mile and veer left onto Wheatland Road, then travel for about 7 miles while following the signs to Wheatland Ferry. The turn for the Willamette Mission boat ramp is just before the ferry landing.
DISTANCE: 12 miles
SKILL LEVEL: Beginner paddlers and up.
CONDITIONS AND EQUIPMENT: The river is mostly wide and gentle, with only Keizer Rapids (no rapids) and just upstream of Willamette Missions providing a few riffles and eddy lines. Good for canoes, kayaks, and drift boats.
AMENITIES: Wallace Marine Park has free day parking, restrooms, and running water. Willamette Mission boat ramp has fee parking and a pit toilet.
WHY THIS TRIP? The wonderful Keizer Rapids Park and Willamette Mission State Park provide an abundance of scenery and wildlife-viewing opportunities. You can also find a couple of nice campsites on this stretch.

The river in this area typically has a gradual current at low to moderate flows. McLane Island appears soon after the put-in and can be navigated to the left or right. This small state-owned property provides a surprising amount of solitude on the channel on river left. After this island, the houses and a few businesses on the right soon transition to River's Edge Park, at RM 82, owned by the city of Keizer.

Below this area, the riverside transitions from willow-covered banks and backyards to a heavily riprapped bank adjacent to the numerous suburban homes. At the point where the riprap and homes end, you have reached Keizer Rapids Park at RM 80. A good stopping point is the gentle eddy at river right. A boat ramp is planned at this site, with adjacent parking.

From Keizer Rapids to Willamette Mission State Park, at RM 72, the river is a bit more quiet and natural. Soon after passing Keizer Rapids you encounter Darrow Rocks on river left, which can create some swirly water. Stay right to avoid the rocks. Shortly after Darrow Bar is Spong's Landing Park on river right and then Spong's Bar at RM 77. The side channel around this island is quiet. If you chose to take this channel, there is ample space to get a canoe or kayak through the middle, but be sure to exercise caution as you approach the pilings at the opening. In this stretch you can see bald eagles, ospreys, and the occasional deer. Such areas have been known to harbor river otters as well.

As you near Willamette Mission State Park, you will see Spring Valley Access on river left at RM 74.5. Shortly afterward, the river crosses a large gravel bar, with half of the river passing over the gravel shelf. If you are in a canoe, kayak, or drift boat, just look for the most promising spot—one is usually as good as the other. The other option is to stay generally left, crossing the bar at the very end, where higher flows can typically be found.

At RM 74, there is a classic opening for a backwater area, part of Windsor Island. This backchannel can be paddled for at least a quarter mile and a range of birds can typically be found here. This is also the historic opening of the channel around Beaver Island at Willamette Mission. New water only enters the channel at higher flows, but this channel could well be reconnected year-round to the backchannel. The area is rather tranquil, other than the sound of gravel extraction that occurs just upstream. As you round past Willamette Mission, there is a relatively high cliff face on the left. The erosive process of the river can be seen along this face, and the house at the top looks perilously close to the edge. To the right the large gravel bar extends downriver, which is the river's main interface with Willamette Mission State Park.

The Wheatland Ferry landing comes into view at RM 72. The ferry provides regular service across the Willamette for a wide variety of vehicles, from the trucks that harvest corn in the summer to a group of Harley riders crossing the river on an afternoon ride. The ferry moves at a surprising speed, so be watchful. Just below the ferry landing is the boat ramp that makes a good take-out. The boat ramp area, owned by the Oregon Parks and Recreation Department, has a pit toilet and day-use parking. It would be great to have overnight parking here in a secure area to serve as a terminus for longer multiple-day trips. ▣

The Willamette Mission State Park is one of the most significant historic sites in the Willamette Valley. In 1834 the Reverend Jason Lee, a Methodist missionary who originally traveled to Oregon with Nathaniel Wyeth's party, constructed a mission high atop the east riverbank along the Willamette River. At this time, the area must have been amazing to see, with the expanse of open prairies meeting backwaters teeming with wildlife.

The mission flourished for more than a decade, serving as a focal point in the sparsely populated Willamette Valley. Members of the mission were later active in the formation of the Oregon government. At the park, the original mission buildings are represented by a framed outline, or ghost structures. Also there is the still-active Wheatland Ferry, the first to carry covered wagons across the Willamette River in 1844. A significant backwater arches from south to northwest within the park. When the mission was established, this backwater was the main river channel. The flood of 1861 shifted the channel a half mile to the west, where it is found today. The mission ghost structures overlook this backwater area.

The park is the centerpiece of the Mission Bottom area, a large bottomland that hosts farms and gravel extraction operations. The Willamette once had an extensive connection to the floodplain lands that surround these bottomlands to the north and south. Willamette Mission State Park provides a good glimpse of this connection, as the park experiences near-annual flooding. At the south edge of the park, the backwater seeps toward the old contoured historic channel at high flows, inundating a good portion of the park in the wettest months of the year. In time, more water may be allowed in the channel year-round. On the main entrance road is a large vertical measuring stick used to measure the flood height, a vivid marker of the park's acceptance of such flooding.

Willamette Mission State Park serves as a reminder of how the river environment can and does change. The park also contains good habitat for a range of species, and several habitat restoration projects dot the property, heightening the awareness that Oregon's parks serve multiple purposes—not simply recreation. The park also provides a glimpse into its agricultural past and present. Much of the park was once a farm, and a good portion of the land nearest the river hosts a large hazelnut orchard which is open to the public for picking in the late summer months. Along the road leading into the park, a portion of the parkland is leased to local farmers who grow berries and other crops.

Just across the river from Willamette Mission State Park is Wheatland, a small town with strong historical roots. It was once a busy stopping point for Willamette River steamboats that made their way up and down the river. Steamboats carried a sizeable amount of cargo before the railroads put these flat-bottomed shallow-draft boats out of business. Prior to the control of the Willamette's flow by the U.S. Army Corp of Engineers, shallow-draft boats were the only way to get up and down the river. Even today, most river traffic above the Yamhill River is conducted by jet boats, canoes, kayaks, and drift boats.

Just past Wheatland Ferry and across from the Willamette Mission boat ramp is Wheatland Bar, a wonderful island owned by the Oregon Parks and Recreation Department and the Department of State Lands. An excursion to this island is only an option for those traveling by river. Although you can look over at Wheatland Bar from the boat ramp at Willamette Mission, there is no road access to the island itself. Nearby is another large public property, Grand Island Access. These areas provide an excellent opportunity to view wildlife, such as wood ducks, green heron, and beaver.

Wood Duck

The wood duck (*Aix sponsa*) can be seen throughout the Willamette system, although this species is not common. If you approach a backwater area, you might see a single bird or a pair zip by as they make a hasty retreat, or you may see one quickly disappear into the brush from the river's edge. The wood duck is one of the most colorful waterfowl that live along the river. The male has distinctive red eyes, a red bill, a crested green head, a white throat patch and finger-like extensions onto the cheek and neck, a chestnut breast, golden flanks, and dark iridescent greenish blue back and wings. Wood ducks nest in tree cavities near the water. When the chicks are ready to leave, they do a freefall from the nest—sometimes from great heights—and head toward the water.

Wood ducks occasionally can be seen along the river, usually in backwater areas.

Wheatland Bar to San Salvador State Park, RM 72 to 57

STARTING POINT: From Interstate 5, take the Keizer exit (260A) and follow Lockhaven Drive to River Road (Highway 219). Head north for about a half mile and veer left onto Wheatland Road, then travel for about 7 miles, following the signs to Wheatland Ferry. The turn for the Willamette Mission boat ramp is just before the ferry landing.

ENDING POINT: From Interstate 5, take Exit 278 and head west on Ehlen Road, which becomes Yergen Road and then McKay Road. Turn left at River Road (Highway 219) and travel 3 miles into downtown St. Paul. Take Blanchet Avenue west to Horseshoe Lake Road, which after 2.5 miles dead ends at San Salvador State Park.

DISTANCE: 15 miles

SKILL LEVEL: Experience with fast-moving water and occasional woody debris.

CONDITIONS AND EQUIPMENT: This stretch is fairly wide and flat, with gentle current in the summer months. Canoes and kayaks work well here.

AMENITIES: Willamette Mission State Park has fee parking and restrooms. San Salvador State Park has parking on gravel lot and a portable toilet. San Salvador is a remote location, so don't leave valuables in your vehicle.

WHY THIS TRIP? It provides a wonderful look at the backchannel of Wheatland Bar, immediately across from the put-in. The backchannel has state-owned lands on both sides. A couple of nice islands can be explored on this stretch. There are several camping opportunities on state lands and a wide slow-moving channel along the last few miles into San Salvador State Park.

Put in at the boat ramp at Willamette Mission and paddle almost immediately across the river to the opening of the side channel at Wheatland Bar. You can enter the backchannel from slightly downstream. At some water levels the few pilings that remain at the opening of this channel can be close to the surface, so be careful as you enter. Once in the channel, the atmosphere changes and the only thing you'll hear is the swirl of the water over rock and riverside branches brushed by the current. Areas like this are wonders to those seeking wildlife, quiet, and relative solitude.

As you come around a slight right bend, the channel to your left is the opening of Lambert Slough. The land between this slough and the main channel is Grand Island. This small branch of the Willamette trails for 10 miles across a lowland area before joining the river again at Lambert Bar, at RM 65. The first mile of land along this slough is public, part of the adjacent Grand Island Access. While the entrance to Lambert Slough has been manipulated over the years, as has the entrance to the side channel of Wheatland Bar, you can get a good sense of how multiple channels interact across a vast floodplain area.

Wheatland Bar is wooded on the periphery, with an open area in the middle. In the spring you can see numerous wildflowers here, especially pockets of camas that dot the interior. As you paddle further down the channel, there is a small island, not more than 20 by 25 feet, that seems to take on the shape of a beaver wading in midstream. Further downstream a large patch of camas can be seen on river right, immediately across from the access point and sign for Grand Island Access. Camping is allowed on Wheatland Bar, though it is rustic with no amenities. Just across from Grand Island Access you can see a wonderful backwater area that has a significant growth of wapato.

As you reenter the main channel, the vista opens with agricultural land to the left and a small wooded area to the right. This is the start of Eldridge Bar Landing, an Oregon Parks and Recreation Department greenway park, at RM 70 to 69. After 4 miles of travel downstream, you can see a large backwater area marking the entrance of Lambert Slough flowing back into the Willamette. This stretch is scenic, though paddlers should be aware of changes from year to year in the small gravel islands that dot the river at RM 64.

At RM 62.5 is Five Island, a small gem in the Willamette River. Five Island has a stellar side channel, with clear and fast-flowing water. Be wary of pilings at the channel opening. Western pearlshell mussels can be found in the backchannel, and bald eagles can sometimes be seen perched in the cottonwoods in this area. Although the entrance to the backchannel has a bit of current, in general the flow is gentle. The upriver end of the island is sandy, and the lower end is characterized by a large patch of wapato that can be seen from late spring through October. Five Island makes a fine getaway, with a rustic campsite that is very quiet.

Candiani Bar is a large island from RM 59 to 58 that is owned by the Department of State Lands. The entrance to the backchannel has a few old pilings, with a clear gap between them, and mild current. Large and rambling, Candiani Bar is a massive gravel bar covered with cottonwoods, and it has a large river rock beach that recedes into the trees. The backchannel is small and winding, but very scenic. This channel is another area where conditions can change from year to year, especially with woody debris, so keep your eyes open.

Back in the main channel after Candiani Bar, the river bends and San Salvador State Park is on the right. This remote 1-acre park was the site of an old ferry landing and today is in the middle of a large agricultural area. Because of its remoteness, at times this park has suffered from vandalism and littering. In recent years the park has seen more use by fishermen and paddlers. At some point, the road leading into this park may closed and used solely as an access point from the river. ▣

Islands like Candiani Bar, owned by the Department of State Lands, can play a significant role as areas of refuge for wildlife and people in the Willamette floodplain corridor. Unlike the Oregon Parks and Recreation Department, the Department of State Lands holds riverlands in trust for the state. The original idea was to issue permits to excavate these islands for gravel, which would in turn provide funds for Oregon schools. However, the Department of State Lands has begun to change its tune in relation to its public lands. Candiani Bar is a perfect example of the potential for conservation and restoration, with its long and scenic backchannel and small backwater area at the downriver end of the island. In such cases, the hope is that the Department of State Lands will designate these lands as conservation areas. A payment from a conservation trust could provide a source of income for schools, while placing the land off limits to resource extraction. Instead, these lands would be left in their natural state or enhanced through habitat restoration projects. For large islands like Candiani Bar, a range of wildlife might find an abundance of good habitat to use, benefiting salmon, Canada geese, and other species.

Wapato at the water's edge in Peoria

Wapato in a backwater at Candiani Bar

Wapato

The large, arrow-shaped leaf of wapato (*Sagittaria latifolia*) is distinctive. This is a native aquatic plant along the Willamette River, especially in the lower reaches. Wapato can be found in backwater areas or where the current is slow and along wet stretches of the riverside. The plant requires a rich muck that is submerged for most or all of the year. For centuries, wapato has been harvested by Native peoples, especially the Chinook, for whom the tubers were a staple of their diet in historic times. In 1806 Lewis and Clark heard reports of Native women wading in water up to their chests or even necks, while using their feet to release wapato tubers from the stems. The tubers floated to the surface, where they were collected and tossed into a special canoe.

In the spring, Canada geese can be seen frequently in backwater areas.

Canada Goose

One of the most commonly seen birds along the Willamette, the Canada goose (*Branta canadensis*) is a large waterfowl with a light grayish body, long black neck and head, and white cheek patches. Canada geese live along lakes and ponds as well, but they are most abundant along the Willamette River, finding refuge in backwater areas and in the Ankeny and William L. Finley National Wildlife Refuges. The birds migrate in the classic V formation and can be seen overhead in the Willamette Valley every spring and autumn. Oregon also has resident birds that stay throughout the year. More and more Canada geese are overwintering in the state rather than continuing south to California, favoring the rich agricultural lands in the Willamette Valley. You may also see Canada geese in local parks or open areas, such as Minto-Brown Island Park in Salem and Oaks Bottom Wildlife Refuge in Portland, but the birds can be particularly wild. Canada geese graze on grasses and similar plants. They also like green forage in the winter months, which has been a source of concern for some farmers and park managers.

Canada geese form a permanent bond with their mates. In the springtime the pairs build nests in locations ranging from backwater areas rich with sand, rocks, and grass to areas as sterile as a cement ledge under a bridge. Their loud honking is especially pronounced in the spring, during the mating and hatchling-raising periods. Soon after the goslings hatch, they are in the water. The transformation from a gosling to a larger subadult bird seems to take place overnight, as the light fluffy down of the goslings begins to turn color. It's also interesting to see the different stages of hatchlings along the Willamette. In a relatively small stretch of river, you may see two adults with goslings that are a few weeks old and then another pair with goslings that are only days old.

About 2.5 miles downstream from Candiani Bar is the confluence of the Yamhill River and the Willamette, with the remaining few miles from the Yamhill to Newberg being part of the Newberg Pool. Compared to the upstream reaches, the current is slower in this stretch. Among the greenway parks and private lands with intact riparian areas, the hum of equipment from gravel extraction can be heard here and there and water withdrawal pipes for irrigation can be seen. On the whole, though, this section provides an expanse of quiet water, with no bridges until you pass the city of Newberg, as well as some hidden treasures.

With a basin covering some 600 square miles, the Yamhill River has its origins in the Coast Range, like the Marys and Luckiamute Rivers before it. The Yamhill flows into the Willamette at RM 55, and the difference between the two rivers is stark. The flow from the Yamhill is brown and can be seen easily, swirling gently with the greener flow of the Willamette, which is many times larger than its small tributary. Prior to the settlement of the state, employees of the Hudson Bay Company called the Yamhill the Yellow River because of its sediment-rich water. When looking at the low flows of the Yamhill in summer and autumn, it's hard to believe that riverboats traveled upriver to take on cargo and passengers in the late 1800s.

The confluence of the Yamhill and Willamette marks the approximate point where the overall gradient of the mainstem Willamette changes. The next 29 miles of the river, down to Willamette Falls, is wide and flat and is generally known as the Newberg Pool. During the summer months, the current slows to one that is almost imperceptible. The formation of the Newberg Pool is due in part to the underlying geology. Here, the river gradient is relatively low. Instead of gravel bars punctuating the river followed by slight

Ash Island in the Newberg Pool

drops where the flow travels down a few inches, the pool area is relatively flat and wide.

After the Yamhill confluence is a wide straightaway that leads to the 100-acre Ash Island at RM 52. Emblematic of what is occurring along the river today, Ash Island is in transition. For many years, the property has been farmed by a private landowner, with a ferry from the east shore providing access for farm equipment. The landowner has voiced interest in selling the property to a public entity as part of the Willamette Greenway Program. The cities of Newberg and Dundee as well as Yamhill County have all expressed interest in such a transaction.

Such purchases can bring up interesting questions of who can have access to the property and for what purposes. With its uniform depth and slow current, this part of the Willamette has been popular with waterskiers, wake boarders, and jet skiers. Activities like these take place all the way to Oregon City, as the density of homes along the river increases dramatically just past Newberg. Keeping properties such as Ash Island in their natural state is an important theme of the Willamette Greenway Program. However, providing access for pleasure boats, horses, bikes, and hiking also have been presented as viable uses for the island. These desires must be balanced with the needs of the river ecosystem.

Paddlers take out at Rogers Landing in Newberg

Because of its significant size, Ash Island can provide excellent opportunities to develop off-channel habitat. This would require some level of excava-

tion of the topography to allow water to more easily move onto the island. Whether wetlands, side channels, or riverside vegetation is developed, areas such as these can provide some very valuable habitat for fish and wildlife. If few amenities are provided for the public, then fewer people will visit Ash Island. This would enhance the island's value for wildlife, as well as providing better aesthetic qualities and opportunities for solitude. As you paddle by this island, consider what its future can be.

Rogers Landing at RM 50.25 is a main access point for boating in the Newberg Pool. On a summer afternoon, numerous ski boats zip up and down the river in this area, just under the shadow of the Virginia Paper Mill.

The city of Newberg is the gateway to the Yamhill Valley, where flourishing vineyards and wineries dot the vast assemblage of south-facing slopes. Although Newberg was not built around the Willamette, in recent years it has sought to better connect to the river. The city center is just over a mile from the river, and its relatively close proximity has made people think about possible links to tourism. A small, efficient cruise boat could leave Portland, travel through the locks at Willamette Falls, and make its way upriver to Newberg, where passengers could catch a bus for a day-long tour of the local wineries. A host of related possibilities exist.

Swallows

From spring through September, one of the easiest birds to see along the Willamette River is the swallow. You may see cliff swallows (*Petrochelidon pyrrhonota*), bank swallows (*Riparia riparia*), and even barn swallows (*Hirundo rustica*) among others flying over the river. Typically seen in groups, these small birds zip along just over the river's surface searching for insects; their quick flight with abrupt changes in direction can make them difficult to view through binoculars. These birds have been known to fly just ahead of a rainstorm moving upriver, making an awesome sight as the wall of rain advances with a wall of swallows just ahead of it.

Cliff swallows nesting under a bridge

7
Newberg to Oregon City
RM 50 to 26

The way of a canoe is the way of the wilderness
and of a freedom almost forgotten.
—SIGURD OLSON

IN SOME WAYS, the landscape along the Willamette from Newberg to Oregon City has not changed as much as other areas upstream. Historically, some of the land along this stretch was open prairie, and today the same land is agricultural. Also, because of the hilly geology in places, the river's course is more constrained here, forming the Newberg Pool. Yet significant changes have been made. With a few exceptions, suburbs sprawl along the river from Newberg to Oregon City, with a wide range of homes from smaller to truly massive. Large green lawns stretch toward the river, with seemingly every house having a dock to access the Willamette. Of course, the Willamette is what provides value to these homes.

One of the main issues affecting this stretch of the river has been a poor reputation related to pollution and the occurrence of skeletal deformities in several native fish species, including the northern pikeminnow (*Ptychocheilus oregonensis*, formerly called the northern squawfish). Beginning in the 1980s, some types of fish were found in relatively high numbers in the Newberg Pool with bent spines, deformed skulls, and malformed fins. Studies conducted by the Oregon Department of Environmental Quality found a much greater prevalence of these deformed fish in the stretch between the Yamhill River and Willamette Falls than in areas further upstream. Additional work in the 1990s confirmed this finding, which prompted public concern. In 2001 the Oregon Legislature eventually provided funding through the Oregon Watershed Enhancement Board to conduct a new study to determine what was causing the skeletal deformities. Given the former levels of pollution in the river in general, the relative lack of flow in the Newberg Pool, and industrial discharges in the area, many people assumed that some form of pollution was causing the deformities in the fish.

The study was led by Oregon State University's Department of Environmental and Molecular Toxicology, with staff well versed in toxic pollution issues related to wildlife. The research team collected numerous samples from

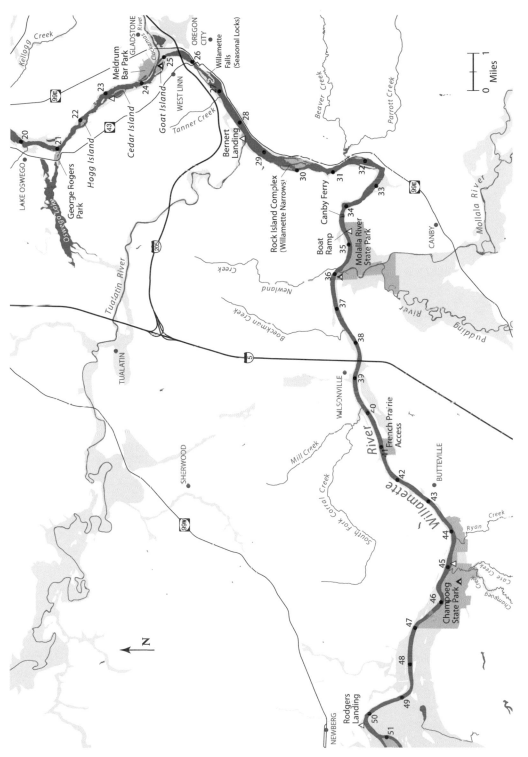

Newberg to Oregon City

throughout the Willamette and found two areas, one in Newberg Pool and the other upriver some 20 miles near Wheatland Ferry, with heightened levels of skeletal deformities. They sought to replicate the deformities in a lab setting, exposing test fish to a variety of contaminants, but to no avail. It seemed that the water chemistry was not the direct cause of the deformities. After months of research, a common connection was finally found among the deformed fish—they all had a fluke parasite. To test whether the parasite was causing the deformities, fish were exposed to the fluke and the overwhelming majority of them developed deformities.

The parasite is a naturally occurring fluke found throughout the West and is transmitted to fish from snails that live in the river. The fluke drills into the bones of newly hatched fish, forming a cyst that disrupts normal bone development. Because minnows with deformed skeletons swim slowly and awkwardly, they are more likely to be eaten by birds, the ultimate host in the parasite's life cycle. Inside the bird, the parasite reproduces and its offspring are excreted, finding their way back into snails to begin the cycle anew.

As part of this research, a northern pikeminnow from the Willamette at the Smithsonian Institution was found to have skeletal deformities, but, curiously, the fish was collected in the 1855. The presence of deformed fish more than 150 years ago helped researchers to see that the problem may not have been from toxic pollutants. While the study has linked the parasite to the fish and the related deformities, what is less clear is why the parasite persists in this area to affect so many fish. Do the parasites occur in greater numbers in the Newberg Pool? Are the fish in this area more susceptible to the parasites, an indicator of weak immune systems? If so, there may be something about the river that is creating part of the problem. For many people, the fact that a parasite was found to be "the" cause of the deformities is good enough, and they see no need to fund additional research into why the parasites are affecting fish in greater numbers in this area. However, others believe that such questions should be answered.

Corvallis withdraws drinking water from the Willamette almost 100 miles upstream, and the city of Wilsonville elected to use the river for its drinking water as well. In 2002 Wilsonville brought its water treatment plant online. Although the plant can treat 15 million gallons a day, as of 2007 the city was using a maximum of 6 million gallons a day. The plant uses standard treatment methods, as well as ozonation and carbon filters. Given the history of skeletal deformities of fish in this stretch and that numerous industrial and municipal facilities discharge their treated wastewater into the Willamette, the public was concerned about the safety of drinking water from the river. However, Wilsonville performs stringent testing of its water system, and the city's treated water far exceeds water quality standards, which is helpful in allaying residents' fears. In fact, using the Willamette River as a drinking water source drives home the idea that making the river's water as pollution-free as possible is a worthy goal.

An issue that has risen to the fore in recent years is the application of mixing zones under the Clean Water Act. The legislation originally sought to eliminate all discharge of waste into U.S. waterways by 1985. Instead, the regulation of point source discharges, which come from a concentrated source such as a pipe or set of pipes, has served as the surrogate for the complete elimination of waste. Some argue that treatment of the waste serves us just fine, even though impure water is still pumped back into rivers. The state of Oregon, which implements the Clean Water Act on behalf of the federal government, has set standards for a variety of pollutants. Water quality standards are set to ensure that pollution levels in treated effluent are safe for the river and its users. Unfortunately, the discharge from point sources does not always meet these standards.

Like many states across the country, Oregon allows mixing zones in the state's rivers. Under the National Pollutant Discharge Elimination System, permits are issued that regulate industrial and municipal discharges into the river, Oregon allows a certain area of a water body close to a point source to violate water quality standards. Typically an oblong area extending up and far down from the discharge point is allowed to violate the carefully developed standards. Officials argue that these areas do little harm to the river and that sufficient dilution has occurred at the edge of these zones. Unfortunately, there are no requirements for the permit holder to test water quality at the edge of this zone. In a river like the Willamette, with large amounts of discharge from both mills and municipal treatment plants, these mixing zones can be large—in one case 70,000 square feet. Extending across a wide expanse of the surface water and downward toward the river bottom, a mixing zone takes up a good portion of the river. The ecological impacts of mixing zones have been a concern, given that fish have to pass through these areas when they migrate.

Mixing zones were clearly not the original intent of the Clean Water Act. However, little research has been conducted to test whether these mixing zones are the source of pollution in the Willamette River. For instance, levels of aluminum and lead in some mixing zones are many times the state water quality standards, yet the dispersal of this material downriver or into nearby sediments has not been evaluated. Although it would be prudent for the Oregon Department of Environmental Quality to take protective steps, the politicians who legislate the agency's budget but who also answer to business interests in the state want absolute proof of harm before any policy changes are made.

This section of the Willamette River is wide and flat, with relatively little current from summer through autumn. Some seek out the area for recreation of the motorized sort, and for them the Newberg Pool, with a uniform depth of 15 to 20 feet and little current, works well. At times in the summer this area is overrun by powerboats, waterskiers, and wake boarders. Many newer ski boats have special devices on the hull that help increase the size of their wakes, so that wake boarders can zip up the large wakes to jump and flip. Homeowners along the river have been affected to one degree or another by wake board boats zipping back and forth, with music blaring. Others prefer to experience the natural wonder of the Willamette River in a paddle craft, enjoying the great blue heron, mergansers, green heron, and other wildlife that use the area. Usually these somewhat competing interests are able to coexist.

The wakes have begun to erode the bank at properties lining the river, affecting trees and shrubs that hug the riverside. Residents along the Newberg Pool voiced their concern and lobbied for several years, until the state began to take some action. In 2007 the Oregon State Marine Board, the entity charged with overseeing the state's motorized boating policies, instated a voluntary slowdown of boats near homes and riverbanks. Coupled with education at boat ramps in the area, the idea was to help wake boarders understand that they need to coexist with homeowners and other river users. If the voluntary effort does not work, the marine board has indicated it will pass a rule outlawing the modifications to boat hulls to increase the size of the wakes.

Some argue that homeowners who live on the river should expect boat wakes and river noise. Most homeowners agree that the river is a public resource, but they have to deal with their docks being in a constant jolting motion all summer and the riverbanks eroding, with trees falling into the river. Such erosion can cause additional turbidity, a measure of water clarity. The noise, typically associated with loud music, is being addressed by increased sheriff's patrols. It would seem that homeowners, wake boarders, and those who travel by paddle craft could find a way of sharing the river that respects the rights of all. An upshot of this small conflict is that all parties involved have recognized their shared interest in clean water, and hopefully this will lead to further actions to ensure river cleanliness and health.

Champoeg State Park (at RM 47 to 43.5) is the site where the first Provisional Government of Oregon was formed. In the early days of the settlement of the Willamette Valley, freemen—those no longer obligated to work for the Hudson Bay Company as trappers and other related personnel—found themselves testing their luck on the wide flat prairies south of the Willamette River. Champoeg was the northern border, and the prairies extended southward for approximately 20 miles, stopping near Lake LaBiche, a natural lake that has now been drained. Some of these prairielands extended to the river's edge in places, providing attractive land to farm, and these areas drew the first of the free trappers.

Natives in the area had used the site for centuries before the arrival of the

Euro-American settlers. The name Champoeg is said to be a Native term for an edible root in the area. Champoeg was near the northern extent of Kalapuyan territory. With its open prairie and proximity to the Willamette River, it made a good meeting place for the Kalapuya and other tribes.

As early as the late 1820s, some former Hudson Bay Company trappers were making their homes near Champoeg. It is said that either Étienne Lucier, Jean Baptiste Desportes McKay, or Joseph Gervais was the first to settle at Champoeg, although their activities probably overlapped to some degree. By the 1830s others had entered the area and began to till the land to establish permanent farms. Soon after, the Methodist missionaries arrived in Oregon, with Jason Lee and others establishing Willamette Mission about 25 miles upstream. This outpost brought additional settlers to the Champoeg prairie and beyond. In 1837 a U.S. Navy purser was dispatched to gather information about the Willamette settlements. He noted thirty male residents at Champoeg; thirteen were Canadians and the others were English or American. Soon afterward more Americans began to arrive, with an influx of more than 100 people at Jason Lee's mission in 1840.

As settlement increased, residents began to desire some form of government in the Oregon territory. Meetings were held, ostensibly to deal with the area's wolves, but these meetings went on to discuss organizing some form of government to deal with other issues, such as estates for which no surviving relatives existed. After much discussion and negotiating about forming a gov-

The wide open river at Champoeg State Park

erning body, a vote was taken at Champoeg on May 2, 1843. After a vote of fifty-two for and fifty against, the Provisional Government of Oregon was formed. Champoeg became a thriving town, with many houses and shops and a variety of people living there. Unfortunately, the town of Champoeg was essentially destroyed by the flood of 1861.

By 1900 the area was owned by two farmers. The last living participant in the vote of 1843 was F. X. Matthieu, who visited the site in May 1900 to inform the Oregon Historical Society Committee of where Joe Meek called for the vote. The farmers agreed to part with the small patch of land and later more, leading to the eventual formation of the Champoeg State Park. Today this spot bears the monument installed in 1901 with Matthieu present.

Champoeg State Park is a wonderful place to visit. The park's small museum offers much to please the history buff, from the story of Native peoples in the region to the early Euro-American settlement of the area. By paddle craft, Champoeg can be accessed just below the monument, while a better access for larger boats is at the dock just downstream. The park has a vast expanse of oak savannah that is home to the native western bluebird (*Sialia mexicana*). Paddlers who seek primitive camping along the river can gain permission from park staff, and a developed campground with yurts is available as well. Although the parkland has been farmed for years, sections are being planted with native grasses and wildflowers to gain back some of the prairie vegetation that likely existed here when the Kalapuya people roamed the valley.

Champoeg State Park

WHERE: Take Exit 278 from Interstate 5 and head west on Ehlen Road, which soon becomes Yergen Road and then McKay Road. Take a right at River Road (Highway 219), and take the next right at Champoeg Road, which leads to the park entrance.

AMENITIES: Historical museum, visitors' center, campground with yurts, restrooms, and drinking water. The park has a $3 day-use fee.

This large park can be enjoyed year-round. Wildlife maintains a presence at Champoeg in all seasons, although certainly spring and summer provide the best birding opportunities. Extensive paved and unpaved hiking trails cover the park. Primitive camping is allowed with permission of the park staff, and the developed campground works well for tents and RVs and has a couple of yurts that can be rented. In the western portion of the park is the historic monument marking the spot where in 1843 the vote was taken to form the Provisional Government of Oregon. From the monument you can walk along the river for approximately a mile to the boat dock. Take the time to visit the Historic Butteville Store founded in 1863 and now run by the Friends of Historic Champoeg, as well as the museum at the park entrance. Here you can learn about the great flood of 1861 and the archaeological findings in the area, among other things. ■

As you travel from Champoeg along the river, at RM 43 you reach Butteville, another historic town and active riverboat port back in the day. Today a few houses are there along with the Historic Butteville Store, which is operated on weekends between Memorial Day and the end of September. From the river all you see is some concrete and a small trail leading upward into the riparian area. Further downstream the riverside is a continuation of suburban neighborhoods interspersed with a few green areas. There is a railroad bridge near RM 39, and then Interstate 5 crosses the river.

The Molalla River State Park begins at RM 36, near the city of Canby at the confluence of the Molalla River and the Willamette. Amid the suburbs, this park provides a good sampling of lowland forest, as it is covered with a sizeable woodland of black cottonwood, ash, and other native trees. Molalla River State Park has walking trails on a large open meadow, a crude boat ramp, parking, and restroom facilities. These amenities are sequestered near the downriver end of the park, leaving the wide expanse where the Molalla meets the Willamette in a natural state. With the confluence area, the park also makes a promising area for research and habitat restoration to benefit species such as spring chinook, western pond turtle, and more.

Molalla River State Park

WHERE: From Highway 99E, turn onto NE Territorial Road on the northeast edge of Canby. After about 1.5 miles, take a right on Holly Street and follow it to the park entrance. Nearby you will also find the Canby Ferry, which provides regular service across the river toward Stafford.
AMENITIES: Restrooms and drinking water.

Molalla River State Park provides an opportunity to walk along the river for about a mile, with an even, unpaved path that descends into the cottonwoods. The peninsula that extends to the confluence of the Molalla and Willamette Rivers is replete with bird life, and the area has the characteristic assemblage of woody debris that indicates its periodic flooding. ■

Just upstream from the park, the Pudding River flows into the Molalla River. In many ways, the Molalla and Pudding Rivers could not be more different from one another. The Molalla flows from its headwaters in the Cascade foothills and makes its rocky descent quickly. With large areas of forest and undeveloped land along its course, the Molalla River has retained a somewhat wild state. The river also flows with a natural hydrograph, leaving the spring rush and summer trickle of its volume to nature. The Pudding River runs slowly across a portion of the valley lowland from Silverton northward, winding its way across a wide agricultural expanse. The Pudding has been altered greatly in areas, with riverside vegetation being denuded. As along the main river, lack of riparian vegetation increases the amount of sunlight that enters the river, resulting in higher than natural temperatures.

As you move downriver from the Molalla River State Park, there are more homes. After making a wide turn to the south at RM 34, a 1.5-mile straightaway ends as the river arches sharply northeast again. At this turn is a very bad example of building along the Willamette River. In the late 1990s, the owner of this partially built mansion excavated some of the river frontage by the house, with no permits to do so. The house is close to the river and easy to see, and the scale of the building is completely out of proportion with other homes along the river. After much public scrutiny, the owner left the building incomplete and moved out of the area, while vowing to someday finish the house or sell the property to someone who will finish it. The average person may wonder why someone would build a home ostensibly to enjoy the Willamette, when it is the house that dominates the river.

Just downriver, lots of basalt becomes apparent along the riverside as you near the Rock Island complex. Rock Island is a large basalt island in the area known as the Willamette Narrows, because the river becomes more constricted as it makes its way through the basalt. The area's unique geology and relatively inaccessible location have also kept it from being developed—no small feat given how close this section of the river is to a large suburban population. This 2-mile stretch of river contains high forested slopes on the west bank and several basalt islands. Here the river is deep, some 80 feet or more in a few places, as it flows through a basalt trench deposited in ancient flows of lava from the Columbia Plateau. Imagining the days of old is also not too difficult here, given that the basalt rock has changed little if any. If you subtract the sound of the nearby highway and the power lines, you can almost see how the Willamette looked 200 years ago.

Metro, the regional government in the Portland area, owns a significant amount of property along the Narrows. On the left shore is Rock Island Landing. With a high sloping expanse of rock and grass, the habitat is perfect for Pacific madrone, which can be seen throughout the Narrows but in high density on this property. A small indentation in the rock creates an embayment on the upriver end of Rock Island Landing, at river left, enabling the river traveler to beach and hike up onto the property. While hikers need be conscious of poison oak, which grows in patches across the hillside, spring brings a colorful blanket of wildflowers interspersed with grasses, making the area a wonderful destination for a few hours exploration via both paddling and walking.

Across the main channel from Rock Island Landing is Rock Island. The island has a cut in its dense Douglas fir and madrone forest for the power lines that cross the river. At higher water levels, you can travel across the river here to access a backwater of Rock Island. If the water is too low, you will have difficulty passing over the basalt and boulders at the backchannel's exit. This passage can be easily seen from the landing.

Downstream on river left is Little Rock Island, an amazing area owned by the Nature Conservancy. This island hosts some elegant native wildflowers, as well as a nice stand of Douglas fir, madrone, and other trees. Little Rock Island also has significant areas of poison oak, so walking its extent can be treacherous. If you take the side channel around Little Rock Island, you will find relative quiet, the sound of the creek rolling down the northern hillside and wind shifting through the trees, with traffic only distantly heard. This short channel has relatively steep banks of basalt, with madrone, fir, and some oaks perched above. Other native shrubs, such as red osier dogwood, grow along the channel. At most flow levels large boulders rest just below the surface at several spots.

What makes the Willamette Narrows fantastic is the pattern of ownership. In addition to Metro and the Nature Conservancy, the state of Oregon owns a large portion of the west bank. At one point, the former owner of Rock Island planned to log it, but Metro purchased the property with funds from a bond measure passed in 1995. Strategic purchase of property adjacent to the Nature Conservancy property and the state land has protected the area. The Narrows has an abundance of madrone trees, native wildflowers, and wildlife, providing the opportunity for native species to flourish in an area close to a large urban center. The success of this effort and others like it was instrumental in helping to pass another Metro bond measure in 2006, with the funds slated to purchase more public lands. With the good will of landowners who support the idea of selling riverside parkland for the benefit of the greater public, there are good opportunities to continue this important effort.

The Tualatin River flows into the Willamette at RM 28.5. Meandering some 70 miles from the coastal foothills in Washington County, the Tualatin is highly impacted by people. With its relatively low flow and gentle current, the river snakes its way across a broad plain, from an agricultural landscape to the thick of the suburbs centered in Washington County. Over the years, the Tualatin was stripped of riparian shade, blocked by debris and construction, and generally relegated to being an afterthought. In recent years, this has begun to change.

Pacific Madrone

Pacific madrone (*Arbutus menziesii*) is a beautiful tree, with smooth, rich orange-red bark that peels away on the mature wood. Although not common along the Willamette, Pacific madrone can be found in abundance at Willamette Narrows and at a few basalt outcroppings downriver. This tree prefers drier rocky areas with coarse soils. With its flat dark green leaves, the tree appears to be deciduous, but Pacific madrone is an evergreen tree. It produces white bell-shaped flowers in the spring that ripen to bright reddish orange fruits in autumn, which are eaten by some bird species. Pacific madrones can become large trees, ranging in height from about 40 to 100 feet under the right conditions. Madrones contrast significantly with the surrounding vegetation and are easy to see at the Narrows and other areas, if you keep a good watch.

Pacific madrone has a red, peeling outer layer of bark and a light, smooth inner layer.

The Tualatin is still beleaguered in places by poor riparian conditions, nonpoint source runoff, and habitat modification, although it does not have large industrial facilities along its banks. Thanks to the efforts of many groups, including small nonprofits like Tualatin Riverkeepers, local landowners, and municipal agencies that treat sewage in the area (in this case Clean Water Services), the Tualatin River has regained some of its natural aspect. During the summer months much of its flow is treated wastewater—which is clean for recreational and wildlife uses—and the water quality and quantity has improved over what existed in decades past. As a key thread for urban wildlife, the Tualatin has increasingly become a valuable natural asset in the Portland metro area that helps to sustain the Willamette's health as well as the livability of the metro area.

Just downriver, reality returns quickly with a view of Interstate 205 perched on the north slope of the Willamette and the hillsides covered with homes. In the distance, the outline of the mills can be seen at Willamette Falls, a 40-foot horseshoe-shaped falls composed of basalt, where the entirety of the river plunges over the chasm. Large pulp and paper mills have long been part of the Willamette's riverscape. At RM 27, the silhouettes of large buildings, with smokestacks and seemingly wayward appendages of steel framing, are seen. Steam emanates from the mill lagoon filled with wastewater that will soon be pumped into the Willamette River. Lagoons help allow the solid

wastes to drop to the bottom, with the clearer effluent on the top to be disposed of in the river or in another pond for additional settling. Viewed from Willamette Falls Drive above the ponds, a lagoon is distinctly whitish blue, a strange and unnatural color for a river. The paper plant that fills this lagoon is a mile downstream.

The Blue Heron Paper Company and West Linn Paper Company have had a mixed relationship with the river and the surrounding area, depending on who you talk to. Although the mills provide a range of good jobs, they have had significant pollution issues over the years. For a time in the late 1990s, Blue Heron was illegally discharging effluent, which resulted in a fine from the Department of Environmental Quality. The company became employee owned in 2000, and the discharge to the river has been improving. This paper mill discharges approximately 8 million gallons of treated effluent per day into the Willamette.

Some people argue that environmental laws hurt industry by increasing costs and making them less competitive. When efforts were going forward in the early 1970s to ensure that mills and municipal facilities treated their waste much more thoroughly, local industrial interests made the same argument. In reality, a major factor that makes U.S. companies less competitive is the very low wages paid in other countries for the manufacture of the same product coupled with few environmental protections. One can hope that employee-

A view looking upriver, with Oregon City in the foreground, Willamette Falls and the adjoining paper and power plants, and the treatment lagoons in the background. Interstate 205 runs parallel to the river.

owned companies like Blue Heron can do the utmost to protect the Willamette River's water quality, while still providing fair wages. Of course, it all comes down to consumers' willingness to pay a bit more for a product that was manufactured in a way that is not harmful to the environment. If someday the mills are forced to move because of global economic conditions, there could be an opportunity to make a state or national park in the area. Willamette Falls are impressive enough to command that type of status.

⚠ You cannot run Willamette Falls in a boat. The only option is to portage from Willamette Park in West Linn or to take the locks on river left. The operation of these locks may be sporadic, so check with the U.S. Army Corps for hours before you seek to boat through this area.

At high water, the roar of Willamette Falls is just shy of deafening, with its massive surges of swirling brown water. Geology also comes into play below Willamette Falls. At the base is an irregular array of basalt formations, which descend downriver toward a deep basalt trench. The depth of the water in this trench can be more than 90 feet near the Oregon City bridge. While Willamette Falls is blessed with natural beauty, it also hosts three major industrial facilities: Blue Heron Paper Company on the east bank, West Linn Paper Company on the west bank, and Portland General Electric's hydropower facility between the two. A fish ladder run by the Oregon Department of Fish and Wild-

At high water, the falls are simply ferocious.

life is nestled against the falls as well. On the east side of the Willamette above the falls, there is viewpoint off of Highway 99E, providing an excellent vista of the falls, the mills, and upriver.

For thousands of years before European-American settlement, the falls were used in a fashion that had far less impact on the river. Chinook, Kalapuya, Clackamas, and other Native peoples once frequented the area, with the impediment of Willamette Falls being a natural meeting place. The Kalapuya brought goods from the valley, such as smoked meat, to trade for smoked fish from the Chinook, who traveled upriver by canoe from Sauvie Island and beyond. The Native peoples met at the falls, portaged around them, and continued up- and downriver as they pleased.

The falls also served as a slowing point for fish moving upriver, with the 40-foot basalt wall providing a natural barrier that stopped most fish. Because the typical spring rush of water poured over the falls in a tremendous volume, the river was high enough to allow spring chinook to swim up rushing chutes of water at the falls to make their way far up the Willamette Valley. With the river's flow now tightly controlled by dams upriver, today salmon and other fish are allowed further passage upstream via fish ladders.

As settlement expanded further into Oregon, a critical issue became how to get around the falls. Portaging was difficult, and carrying both cargo and craft over the falls was time consuming and dangerous. In 1850 a half-mile portage road was constructed on the east side of the falls, running from Canemah to just below the falls. Freight was hauled at first on horse-drawn carts and later on horse-drawn rail. A breakwater extending to the east side of the falls was later created. An adjacent stairway led down to the lower river level, where passengers could board another boat to continue their voyage downriver, and an elevator carried cargo down to the lower river level. Finally, in

Pacific Lamprey

Spring chinook are not the only native aquatic species of interest at Willamette Falls. Pacific lamprey (*Lampetra tridentata*) have had a long history with the falls and the Chinook, who harvested these primitive fish. Lamprey can be quite tasty when cooked over an open fire.

Lamprey are slimy to the touch, thin, and long, reaching up to 30 inches in length. They are jawless fish with sucker mouths. As adults, lamprey are parasitic and attach themselves to the side or undersurface of other fish from which they draw blood and body fluids as food. Lamprey spend their juvenile years in the upper tributaries of the Willamette Basin and then travel downriver to the ocean over a period of years. They spend a good portion of their life in the Pacific before returning upriver to spawn—many at Willamette Falls.

A large group of paddlers from Paddle Oregon head downriver in the Willamette Falls Locks.

1873 a system of locks was constructed that could float craft up and over the falls and back down. Railroads soon made significant commerce via riverboats a thing of the past, and the last steamer along the Willamette ceased operations in 1940. After that time, the locks were used commercially to float log rafts for use at the West Linn Paper Company. Today, however, raw material is shipped to the plant by trucks.

The Willamette Locks are the oldest continuously operating, multi-chambered locks in the United States, although they are a bit of a relic in modern times. If you take a boat up or down these locks, their narrow and relatively deep chambers smell of old wet wood, and small ferns grow in the recesses of the rock walls. This trip is well worth the time, though, as it provides a ready way to get a sense of this area's industrial past and present. Since 2000 the U.S. Army Corps has had their budget for operation of the locks diminished, with the potential to have them closed permanently. Clackamas County, Oregon City, West Linn, and others have worked to keep the locks open and are exploring creative options to keep them open on a regular basis, if even only a couple of days a week or by special arrangement.

Adjacent to the locks is the T. W. Sullivan power plant operated by Portland General Electric. This is one of the oldest operating hydropower plants in the United States. The first generation and transmission at this site dates from 1889, when 14 miles of transmission lines went from the falls to Portland, lighting streetlights in the early city center. At the upper portion of the falls, a retaining wall creates a uniform water line as the water rushes downward. During the summer months, the wall is made taller to direct flows into the turbines at the Sullivan plant.

The Federal Energy Regulatory Commission oversees the permit for the power plant. In 2000 the plant underwent a relicensing, which, amazingly, occurs only once every 50 years. The commission directed Portland General Electric to develop improved fish ladders to better enable both adult and juvenile fish to pass over the falls. The ladders are specifically targeted toward spring chinook. With the falls being deprived of the historic high spring flows, coupled with the intensive industrialization in the area, native spring chinook are no longer able to pass over the falls in gushing chutes of water. Instead, they must use the fish ladders.

Situated at the base of Willamette Falls, the ladders are built under the water flow on the West Linn side. The fish swim through several openings at the base of the falls, and the ladder condenses to one main chute. A small office operated by the Oregon Department of Fish and Wildlife has a viewing window into the greenish water, which enables technicians to count the spring chinook as they pass upstream. Each year the count is critical in establishing the total run of adult fish, both native and hatchery-reared stock.

In the springtime, you can witness an old rite related to the surge of fish up the Willamette in this area, California sea lions (*Zalophus californianus*) feasting on spring chinook. Some argue that the arrival of sea lions is a more recent phenomenon, a result of fish being stalled at the ladders and other food sources downstream leading to higher numbers of sea lions in the river. True or not, today some California sea lions routinely spend a good portion of April and May at the falls and throughout the Lower Willamette. From a canoe or kayak, the sudden appearance of the large mammals rising briefly to peek and puff at unwary paddlers can be unnerving. But sea lions are amazing to experience up close, and they provide some further insight into the natural world of the river. At the falls they are able to rise and snatch up the chinook. The good news is that only a few sea lions are typically found here, perhaps because of some balance with the amount of fish available or the long distance from the Columbia (26 miles) and the ocean (some 100 miles). Fishermen have long been irritated by the sea lions, given that they can present some intense competition for chinook in certain areas at the height of the spring migration, but their anger is misplaced. Destruction of habitat and major modifications to the Willamette system, such as the dams operated by the U.S. Army Corps, have eliminated tens of thousands of these native fish.

A view looking northward from Willamette Falls (c. 1895). Note the power lines than run from the hydropower dam to downtown Portland in the distance. Drawing by Frederick A. Routledge. Oregon Historical Society, no. OrHi48110

8
Oregon City to Sellwood
RM 26 to 17

*Our eventual aim is to divert all wastes from the river . . .
and I'm convinced that it can be done.*
—L. B. DAY

Statue of John
McLoughlin at
the Highway 99E
overlook of
Willamette Falls

OREGON CITY was settled in 1829 and served as the focal point for early commerce, industry, and politics in the Oregon Territory. The city only had a few structures before the 1840s, with most of the residents stationed there from the Hudson Bay Company to create and service a saw mill at the falls. In 1842 John McLoughlin, Chief Factor of the Hudson Bay Company, platted and named the establishment Oregon City. In 1843 the Provisional Government of Oregon was formed, and Oregon City grew as more people poured into the region, fueled by the expanding Methodist mission of Jason Lee. The city incorporated in 1844 and became the seat of the new Provisional Government the following year. By the time Oregon became a state in 1859, Oregon City had hundreds of residents.

Willamette Falls has always been a natural place for people to stop and a popular fishing spot, from the early Native peoples who utilized the falls area for spring chinook and lamprey to those who today fish off the high basalt bluff for sturgeon. Although the spring chinook runs are greatly diminished compared to historic runs, today the run of hatchery spring chinook remains popular. A great many boats travel the area, with some using hog lines in which several boats tie up next to one another.

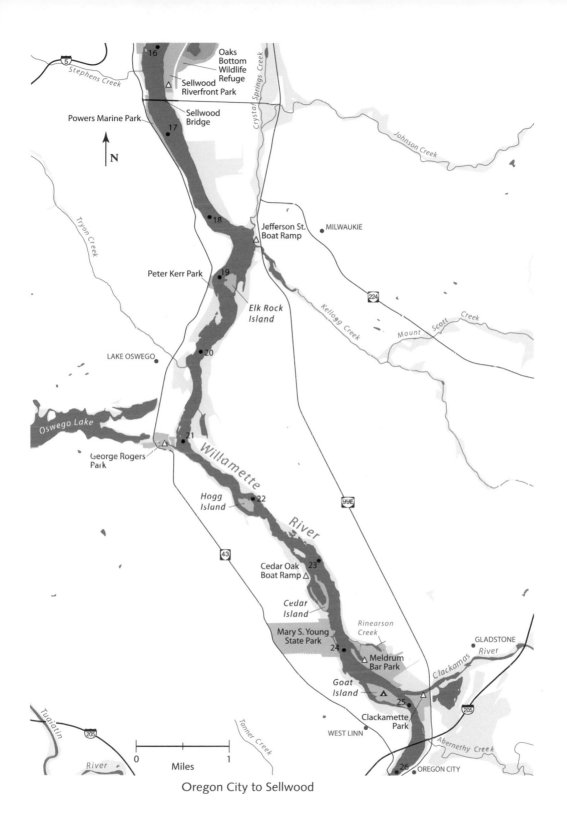

16

Oaks
Bottom
Wildlife
Refuge

Sellwood
Riverfront Park

Stephens Creek

I-5

Crystal Springs Creek

MILWAUKIE

Johnson Creek

Sellwood
Bridge

Powers Marine Park

17

N

Tryon Creek

18

Jefferson St.
Boat Ramp

224

Peter Kerr Park

19

Kellogg Creek

Elk Rock
Island

Mount Scott Creek

LAKE OSWEGO

20

Oswego Lake

21

George Rogers
Park

Willamette

99E

Hogg
Island

22

River

43

Cedar Oak
Boat Ramp

23

Cedar
Island

Rinearson
Creek

GLADSTONE

Clackamas River

Mary S. Young
State Park

24

Meldrum
Bar Park

205

Goat
Island

25

Tualatin

I-205

Tanner Creek

Clackamette
Park

River

WEST LINN

Abernethy Creek

0   Miles   1

26   OREGON CITY

Oregon City to Sellwood

FISHING FOR SALMON, WILLAMETTE RIVER, NEAR PORTLAND, OREGON.

A hog line of salmon fishing boats near where today the Interstate-205 bridge crosses the Willamette at Oregon City (c. 1930). Wesley Andrews photograph, Oregon Historical Society, no. OrHi35424

Oregon City was established because the river and falls dominate the natural setting. Over the years the power of Willamette Falls has been used to run saw mills, paper mills, and a hydropower plant. Because the river was captured by industrial sources early on, in Oregon City the Willamette was seen more as a thing of utility rather than a natural resource to be protected. In more recent times, however, the city has sought to better connect with the river, like so many other cities and towns in the Willamette Valley. Although the area continues to be dominated by the two paper mills and the Portland General Electric hydropower facility, Oregon City has an excellent opportunity to get people closer to the river with the development of new parks and boat docks. Today, river access is easily available only at the boat ramp at Sportcraft Marina and a floating dock just downstream of the Interstate 205 bridge. This marina has been one of the best canoe shops in the area, and its excellent access to the river made it easy for people to engage in a quiet mode of river travel. Ironically, the marina is being displaced by the city's effort to reconnect to the river. In 2009 the shop will relocate further north on McLoughlin Boulevard.

Willamette Falls from Sportcraft, RM 24 to 26.5

STARTING AND ENDING POINT: Boat ramp at Sportcraft, 1701 Clackamette Drive, Oregon City. You can also access the river from Clackamette Park, just downstream of Sportcraft.

DISTANCE: 5 miles

SKILL LEVEL: Experience in mild current and boat wakes.

CONDITIONS AND EQUIPMENT: Good for canoes and kayaks at low summer flows. There is relatively little current along this stretch and only a small amount from the falls. Paddlers must be aware of powerboats and wakes rebounding off hard surfaces, like cement revetments.

AMENITIES: There is parking at both Sportcraft and Clackamette Park, with a portable toilet at Sportcraft and restrooms at Clackamette.

WHY THIS TRIP? This short trip enables you to see the connection between industry and the Willamette and to view petroglyphs and the historic Willamette Falls Locks.

If you stay on the Oregon City (left) side of the river as you head upriver toward the falls, you may be able to see some petroglyphs on the rocky riverside. Almost immediately adjacent to the last of the rocks that sticks out, before you enter the more turbulent water of the falls, look upward to the left. On a rock

Looking downriver under the historic Oregon City Bridge, with the Interstate 205 bridge in the background

very near the river, you can see several rounded figures that have been carved into the stone. The spot, extending outward over a deep-water area, would have been a promising one for the Natives who fished the area for generations. Today the property is owned by the Blue Heron Paper Company, so you should not trespass. Even if the property were public, people should not get too close to this fragile ancient artwork.

Framing the river on both sides in the area is lots of basalt, with high ledges at Oregon City. In general, the area is not too inviting with the Interstate 205 bridge, mills, and abundant traffic noise, but on a summer day you can get a great sense of this place with good access to the falls at low water levels. Rocks appear at low water on the west side of the river next to the West Linn Paper Company. Here it is possible to imagine the scene before the mills were constructed, with Chinook canoes in the area and people fishing along the falls and hiking over a rocky path to meet a group above.

By heading downriver from Sportcraft, you will reach the confluence the Clackamas River, with its cold water mixing with the deeper green and brown of the Willamette. The inflow from the Clackamas is vibrant in spring, and when combined with the mainstem the water can almost burst downriver. The Willamette is fairly shallow at the confluence where rocky detritus tumbles in from its lowest major tributary. The Clackamas Basin is large, nearly 1000 square miles, and its highest tributaries at Olallie Butte are fed by snowmelt from the Cascades.

Directly across from the Clackamas confluence is the upriver portion of Goat Island, a gravel bar island covered with black cottonwoods. Like some other gravel bars, Goat Island is owned by the Department of State Lands. Although it is within range of the rumble of the Interstate 205 bridge, Goat Island hosts a surprisingly large group of great blue herons. In the wintertime, the large rookery can easily be seen in the tops of the trees, usually toward the western edge of the island. The heron rookery is best viewed from the little backchannel found there.

An old tire, too often seen along the river, and the tracks of a deer on Goat Island

Clackamette Park is nestled between the Clackamas and Willamette Rivers on the east bank and provides good access to Goat Island. This park remains one of the best areas to get a sense of the raw power of the rivers in the winter and early spring, when their flows are very high, with the rush of the falls just 1.5 miles upriver. Clackamette Park usually gets inundated with water even at moderate flood levels, with large logs and

other woody debris whirling in the big muddy current. Meldrum Bar Park be- | The flooded
gins just north of the Clackamas River confluence. At RM 24, a large gravel bar | Clackamette
in the park runs parallel with the main current. ▣ | Park in 2006

Part of the greater Gladstone watershed, Rinearson Creek is a small ephemeral creek that winds its way through suburban neighborhoods, crosses under a major thoroughfare, traces its way under an auto dealership, and then enters the backwater at Meldrum Bar. This small Willamette tributary appears to be pretty far gone, surrounded by development and with little in the way of native habitat. Yet some people recognized that this creek could have cleaner water and provide benefits to wildlife.

In 2006 several neighbors got together to investigate what they could do for Rinearson Creek. Over a year and a half, the volunteer-led group, assisted by Willamette Riverkeeper and others, made excellent progress by removing invasive species, such as Himalayan blackberry and English ivy, and planting native vegetation that is more beneficial for native fish and birds along the bank. Small creeks like Rinearson can provide a brief refuge for salmon and other fish making their way down the Willamette River. This effort is a good example of local people taking matters into their own hands and fostering community-based restoration work.

Mary S. Young Park, a state park managed by the city of West Linn, is located at RM 24 on the west side of the river. Young donated the land to the state of Oregon in 1955, explicitly stating that it should never be sold or developed and should be kept in manner that enhances the park's natural attributes. The entrance to this large park is on Highway 43 in West Linn. A paved hiking trail transitions to gravel and makes its way downhill to the scenic Willamette's shoreline. The beaches are separated by basalt extrusions that provide good river views. From the park, you can walk along the beach, heading downriver, to a footbridge that extends over to Cedar Island, a 10-acre island owned by the city of West Linn. For those traveling by paddle craft, the small backchannel of the island is passable except at low tide during low summer flows.

Cedar Island is a shell of the former island, which was dredged out for gravel aggregate in the 1950s. The shoreline is overgrown with Himalayan blackberry in many places. Cedar Island has a crude path and several platforms that extend over an artificial lagoon. This shallow-water habitat in the center of the island is best suited for smallmouth bass and native fish such as largescale sucker. If revegetated with native plants, Cedar Island could provide some important habitat for native fish and other wildlife.

Cedar Island offers something special in this suburban stretch of the Willamette: a group of resident beavers may be seen along the shoreline, swimming in the water, or sedately grooming themselves at the water's edge. While tame by no means, some members of this small unit are acclimated enough to let you pass nearby in a canoe or kayak without running for the water or diving under the surface. Like other areas of the river, beavers have adapted to some degree to the presence of people. Seeing them making their home at Cedar Island may help people to build a healthy respect for these large, gregarious rodents. The lagoon and backwater can be accessed at the Cedar Island boat ramp as well, providing a view of Cedar Island and the lagoon.

Oregon grape can sometimes be found under the canopy along the river.

Hogg Island, at RM 22, is a bit different from many other islands in the Willamette River. This 10-acre island is supported by a foundation of basalt rock, making it sit higher in the water than islands composed of river rock. Hogg Island is owned by Clackamas County and for years has been placed in their "surplus land" category. Ironically, this surplus land is very nice and anything but run of the mill. Although Hogg Island has its share of invasive species, there is a large expanse of Oregon white oaks, snowberry, Oregon grape (*Mahonia aquifolium*), and nice patches of native camas in the spring. For a place located in the middle of suburbia, Hogg Island is surprisingly tranquil.

In 2005 Willamette Riverkeeper began working with Clackamas County to remove some of the invasive species on the island, such as Scotch broom and Himalayan blackberry, which covered a significant expanse of the areas not harboring natives. The clearing revealed native species, such as camas, in some areas. Hogg Island is not an official natural area, but it should be. Oregon white oak habitat has been dramatically reduced in the Willamette Valley, and although Hogg Island does not harbor typical oak savannah, it is worthy all the same.

Perhaps because of its rocky shoreline, except for one easy access point on the beach along the island's backchannel, Hogg Island has been relatively untrammeled by people. Another saving grace may be the poison oak interspersed with other plants along the trail and throughout the island. The poison oak can be extremely pervasive, winding its way along the trail, from ankle level to high overhead. If you have even a slight allergy to this native species, you may wish to stay in the open area above the beach. Unfortunately, this area is often littered with a lot of trash, especially around summer holidays. Some people seem to have little respect for nature, while others have learned the ethic of leaving no trace—in essence leaving natural areas better than you found them by cleaning up litter and leaving nothing behind.

From the small open area on Hogg Island, a path leads to an opening amid the oaks and Oregon grape. Traveling further down the path, to the left you can see the vegetation transition to maples and ash. The main-channel side of Hogg Island is more open, with grass and scrubby Scotch broom that is slated for eventual eradication. This area has harbored some of the recovered camas,

Scotch broom has seed pods that can last decades, making this invasive weed very persistent.

whose flowers grew up after the Scotch broom was removed. The circular path then winds back to the oaks and the open area. Hogg Island's natural attributes can be explored and enjoyed year-round. Despite the occasional sound of traffic on Old River Drive, the secluded forested landscape of the island provides a nice respite from the dominant view of the Lower Willamette River. Making a stop here can afford an opportunity to see native vegetation, birds, and perhaps deer, which swim across even the deeper portions of the Willamette to graze on

island vegetation. It is possible that deer reside on Hogg Island full time, although they only occasionally present themselves.

Just across from Hogg Island is a steep bank of basalt rock, forming a wall that extends nearly a half mile downriver. One of the major property owners here is Marylhurst University. In the late nineteenth century much of the riverside in the area was denuded. A small railroad made its way along the shoreline, and then up the ravine near RM 21. Today Old River Drive on the crest of the basalt layer follows the former path of the railroad and then runs parallel to the river. The high rocks along the shoreline provide a great fishing spot, perhaps for the sturgeon that make their home in the depths of this basalt ravine. As you go downriver from Hogg Island, the depth drops from about 40 feet to more than 100 feet in places near George Rogers Park, at RM 21, where the river turns sharply and heads in a more northward direction.

The smokestack from an old iron smelter can be seen through the trees at George Rogers Park. This chimney structure is all that remains of a significant complex that operated from about 1865 to 1872 on this flat peninsula next to the river. The smelter made iron from ore derived from the hills near Lake Oswego and later Tryon Creek. George Rogers Park is also the place where Oswego Creek enters the Willamette, carrying a portion of Oswego Lake's water into the river. A small hydropower operation manages the flow from

A photograph from about 1900 of where Sucker Creek, today's Oswego Creek, enters the Willamette. Note the cattle grazing on the beach and what appear to be drying fishing nets. Oregon Historical Society, no. OrHi88765

Lake Oswego, a convenient arrangement that allows the lake's height to be managed for the homeowners in the area.

As in so many areas throughout the Portland metro area, houses have filled in most of the landscape, and lawns and manicured yards define the river's edge. Watershed managers and other interested parties have begun to consider how the yards along creeks and other rivers in the basin are affecting the Willamette River. When you look at the homes at the river's edge and the myriad houses beyond along every ravine, meadow, and valley in the area, it's not hard to imagine the volume of pesticides and other chemicals being used to maintain these yards, gardens, and landscaping. These chemicals affect not only the weeds at the target site, but once they are washed into the water system, the chemicals can affect fish, other wildlife, and their habitats.

The idea of landscaping with nature in mind, often referred to in the Portland area as *naturescaping*, has resonated with people for years. The city of Portland and Metro helped to pioneer this notion, and the idea has caught on throughout the Willamette Valley. Willamette Riverkeeper produced a guide to riverscaping in 2002. Naturescaping aims to reduce the use of chemical fertilizers, pesticides, herbicides, and water resources. Pulling weeds, instead of spraying them with chemicals, has been encouraged along with the greater use of species that are native to the region. Native plants take less time and

The rustic east bank of the Willamette at Oak Grove, just across from Lake Oswego (c. early 1920s). Today this stretch of the riverside is covered with houses and condos. Gifford & Prentiss photograph, Oregon Historical Society, no. OrHi105994

resources, because they are adapted to the Willamette Valley's environment and don't need fertilizers, pesticides, and lots of water to do well. It also makes sense to reduce the size of landscaping features such as lawns, which require a lot of water. In spite of the greater availability of native plants at area nurseries and the numerous hours spent pulling weeds to avoid spraying gallons of weed killer, there is still much to be done in reducing the use of water, pesticides, and herbicides at most homes.

As you travel this section of river, though, you'll see large sloping yards that have left many native trees in place, with relatively small lawns and ample use of native shrubs that love the rainy, cloud-filled weather and survive the brief summer drought with ease. The good thing about naturescaping is that you can quickly begin to modify how you care for your yard. In a short period of time, a house with a long green lawn that slopes toward the river can be transformed into a yard with Oregon grape, gravels, and wood. Whether your house is next to the Willamette, a local creek, or in the middle of Northeast Portland, everyone is capable of reducing runoff and the use of chemicals and providing plants that can attract native wildlife. For more information, contact Willamette Riverkeeper (see Resources).

A good example of how yards and gardens can impact water quality is Oswego Lake, which drains into the Willamette via Oswego Creek. This lake, an expanded version of the former Sucker Lake, has had serious problems with summer algal blooms and aquatic weeds. The vivid bluish algal growth does little to invite people to get out on the lake. Oswego Lake also has little in the way of natural shoreline or other features that encourage native species. Most of these water-quality problems are related to phosphorous, the major component of fertilizers once used regularly in the homes that line nearly every foot of the lakeshore. Recently, the Lake Oswego Corporation and the city of Lake Oswego have sought alternatives to phosphorous-based fertilizers, which seem to have improved the situation to some degree. Over the years, the city used copper sulfate to help kill the algae blooms, and excessive levels of copper have been found in the sediment in the lake bottom. It makes one wonder what comes out into the Willamette from Oswego Creek. Perhaps as the Lake Oswego Corporation and the city continues to grapple with the algae issue, they can also focus on helping the lake to function in a more natural manner, which will result in better water quality.

Tryon Creek enters the Willamette's west bank near RM 20. The small confluence area can be a bit hard to see from midriver, but it's easy to spot if you hold close to the left shore. Tryon Creek drains a watershed that has an abundance of suburbs as well as a large state park with a substantial amount of healthy and improving habitat, including Douglas fir forest interspersed with a variety of other native trees. Tryon Creek can provide real benefits for fish and other wildlife.

Fishing Along the Willamette

For as long as humans have lived near the Willamette, they have fished it. The Chinook and Kalapuya peoples took lamprey, spring chinook, sturgeon, steelhead, bull trout, and other fish from the river. Settlers learned how to harness these seasonal gifts and passed on this knowledge to later generations.

Today, many of the fish in the Willamette are introduced species. Smallmouth bass (*Micropterus dolomieu*) is one of the most prevalent introduced fish sought along the river. The best habitat for bass is shallow areas, where there is less current. Black crappie (*Pomoxis nigromaculatus*) are also popular introduced game fish that can be found throughout the Willamette system. The introduced common carp (*Cyprinus carpio*) now occupies much of the river system as well. This extremely adaptable fish eats almost anything and has destroyed significant shallow-water habitat along the river.

The native species sought by fishermen include spring chinook (*Oncorhynchus tshawytscha*), coho (*Oncorhynchus kisutch*), steelhead or rainbow trout (*Oncorhynchus mykiss*), cutthroat trout (*Oncorhynchus clarki*), white sturgeon (*Acipenser transmontanus*), and Pacific lamprey (*Lampetra tridentata*). Historically, only winter steelhead, spring chinook, and lamprey migrated annually upriver of Willamette Falls. Winter steelhead travel above the falls from February through May, and summer steelhead are now stocked above the falls and provide sport fishing opportunities. Coho are mainly fished in the Clackamas River system.

A short hog line just below the Oregon City Bridge during spring chinook season

From Eugene to Harrisburg, you can find native cutthroat trout in the shallow stretches of the river, zipping up through the gravel beds and riffles. This is a prized fish for those fly-fishing the Willamette, although at present cutthroat can only be caught and released. The native rainbow trout can be caught throughout the Willamette system, but these fish are most often found in the cold clean waters of the Upper Willamette.

Spring chinook represent the pinnacle of what can be caught in the Willamette system. It's not uncommon for a chinook to weigh 30 pounds. These prized fish are sought by fishermen from late winter through late spring, before the salmon migrate up through Willamette Falls from March through July to spawn in their natal streams. Prior to the construction of dams in the Willamette system, spring chinook runs numbered in the hundreds of thousands. Today most of the chinook that return to the Willamette from the Pacific Ocean each spring are hatchery-raised fish. Common estimates are that 5 to 15 percent of the total chinook population is naturally reproducing. Each year, about 8 million smolts are raised in four large hatcheries, as well as fingerlings (more mature fish) that are dispersed in streams and reservoirs.

The spring chinook run varies greatly from year to year, and the run is not always large enough to allow for sport fishing. As the salmon pass through the ladders at Willamette Falls, they are counted. The spring chinook run was 85,000 in 2003, 143,700 in 2004, 116,900 in 2005, 59,700 in 2006, and 40,500 in 2007. Sometimes as early as February, the fishing boats begin to appear at the Sellwood Bridge, at the confluence of the Clackamas and Willamette Rivers, and near the Oregon City bridge. In those years with a large run, there are literally hundreds of fishing boats between Sellwood and Oregon City. Hog lines, a series of boats tied one next to the other across a segment of the river, are also common. These hog lines create a weaving array of baited hooks for the fish to pass through as they make their way upstream. Only hatchery-bred chinook can be kept by fishermen. These fish can easily be distinguished from the native salmon by the fact that their adipose fin—the fin just above the tail—is clipped before they leave the hatchery. Any chinook that has an adipose fin must be released.

Another significant native fish of the Willamette is the white sturgeon, which is most often sought on the Lower Willamette, in the deepest reaches of the river. These massive fish live on the river bottom and can reach 75 years in age, more than 10 feet in length and up to 1000 pounds. Just below Willamette Falls, near Elk Rock Island, and Portland Harbor are popular spots for sturgeon fishing, given that these areas are deep as compared with the rest of the river. Sturgeon sometimes breach the river's surface, and it can be a big surprise to see such a huge fish surface and then splash down.

Opportunities to fish on the Willamette and its tributaries exist throughout the year. The numerous habitat restoration projects along the river are
aimed at improving conditions so that healthy native fish populations can be
maintained. You can find out about Willamette River fishing regulations and
licenses from the Oregon Department of Fish and Wildlife. There are also
many guide services for fly-fishing on the Upper Willamette and its major
tributaries, as well as local clubs that highlight fishing and conservation, such
as the Northwest Steelheaders and the McKenzie River Flyfishers.

Since the early 1990s, the city of Portland, Oregon Parks and Recreation
Department, and Tryon Creek Watershed Council have invested significantly
to restore the creek, from its headwater area in the west hills of Portland to
Tryon Creek State Park to the confluence with the Willamette. The aim of the
large riparian restoration project at the confluence is to provide some refuge
areas, with food and cool water, for native fish migrating downriver, including adult spring chinook, as well as juvenile fish in the area. Improving the
water quality has been another goal, as much of what surrounds the watershed is suburban homes. Additional effort has gone into amending the structure of the stream by adding woody debris and making changes in the culvert
running under Highway 43. With such improvements, more native fish can
find their way into Tryon Creek and other similar creeks in urban areas.

Just downriver from Tryon Creek at RM 19 is Elk Rock Island and Peter
Kerr Park on the west bank. In 1910 the island was acquired by Peter Kerr, who

Looking upriver
toward Elk Rock
Island from the
Jefferson Street
Boat Ramp in
Milwaukie

used it for recreational purposes. He gave the property to the city of Portland in 1940 and stipulated that it be preserved for public enjoyment. This large basalt island is a peninsula at low flows, with rock connecting Spring Park on the east bank to the island proper. Elk Rock Island arose from a lava flow that occurred some 40 million years ago, which is distinct from the usual mass of 15- to 20- million-year-old Columbia River basalts that dominate the area, making this some of the oldest exposed rock in the area.

Elk Rock Island has a variety of microhabitats. On the northwest point is a small collection of Oregon white oaks, and to the east the forest transitions into Douglas fir and a few cedars. The south end of the island is replete with Pacific madrone. Among the low-lying rocks are small pools with algal blooms. A path traverses the interior of the island, providing good views throughout. A bald eagle nest has been built just to the east of the island in Spring Park. The proximity of this nest to homes is just amazing, as the birds are literally next to several suburban residences. The great birds can be seen frequently in the area, adding to the wild essence of this island park.

Across from the island is Elk Rock, a relatively steep cliff face that extends upward to Macadam Avenue. In 2007 a pair of peregrine falcons (*Falco peregrinus*) nested on the cliff face, reinhabiting their historic territory for the first time in decades. With a spotting scope and a little luck, the nest can be seen from Elk Rock Island. Key times are late spring through midsummer, when fledglings might be spotted. While nesting habitats of peregrines can change, nearby residents hope that the falcons will live along this cliff face for some time to come.

Ledges along the face of Elk Rock make excellent nesting areas for peregrines. Given that pairs have nested successfully on some of the region's bridges in recent years, some bird experts have wondered when they would regain their natural habitats in the Portland area. The Audubon Society of Portland, in partnership with the Oregon Zoo, Oregon Department of Fish and Wildlife, and Oregon Department of Transportation, has done excellent work to ensure that peregrine falcons have a fighting chance in the area. Many of the region's bridges are scheduled for improvement during the next few years. Although the goal is always to allow nesting peregrines to complete their nesting cycle without disturbance, Audubon and the state agencies recognized that disturbance would be inevitable in some cases. In 2001 a plan was developed to protect these peregrines and their offspring by removing eggs or nestlings from the bridges slated for construction and raising them in captivity. Thus far, five peregrine falcons have been successfully released at Ridgefield National Wildlife Refuge in southwestern Washington.

In the early days of the Oregon Territory, Milwaukie was one of several small towns that vied to be "the" city along the Willamette River, each making claims of having the best port. Like Oregon City upriver, residents of Milwaukie sought to capture the growing river trade, making many statements about its place as the "head of navigation." The founder of Milwaukie, Lot Whitcomb, had a large riverboat named after himself that was part of the early riverboat trade along the river. Today, the Kellog Treatment Plant dominates the Milwaukie shoreline, with the Jefferson Street boat ramp just downstream. The city has made many improvements along McLoughlin Boulevard, to be accompanied by upgrades to the Jefferson Street boat ramp and Milwaukie Riverfront Park.

Johnson Creek flows into the Willamette at Milwaukie. The Johnson Creek Basin is 54 square miles, with about 50 percent of the watershed residential, 8 percent commercial and industrial, 33 percent rural, and the rest in parks and open spaces. The flow in Johnson Creek is very low during the summer months. Historically thousands of salmon found their way into this small watershed, whereas today there are but a few. Although Johnson Creek is affected by problems including E. coli bacteria, DDT, and high temperatures, it represents another good opportunity to restore fish habitat in an urban area.

Hemmed in by highways, parking lots, residential streets, and thousands of homes, Johnson Creek has a long way to go in terms of functioning more naturally and providing clean cold water for fish. Redefining the relationship between people and the creek has been critical. Portland, Milwaukie, the Johnson Creek Watershed Council, and many residents and businesses have worked to curb chemical dumping, buffer the creek from parking lots and lawns, and encourage less use of pesticides for yard and garden maintenance in the watershed. Like the Willamette, having functioning floodplains and riverside vegetation is critical for creeks as well. Toward this aim, the city of Portland purchased property along Johnson Creek, removed a home that was in the floodplain, and developed a floodplain restoration project where the house once stood. A large part of what still needs to be done is the continued education of landowners in the Johnson Creek Basin. Even today, people hack away vegetation along the creek and allow runoff from all manner of properties to get to the creek.

From Milwaukie, a 2-mile straightaway leads past a nice beach at Powers Marine Park to the Sellwood Bridge. A ferry once operated where the bridge is now, with access from Spokane Street on the east side and the boat ramp at Staff Jennings Marina on the west. The Sellwood Bridge, a narrow two-lane span constructed in 1920, carries a lot of traffic each day. In 2002 a weight limit was placed on the bridge, allowing only cars and light trucks to travel across. A new bridge is likely to be built just downriver, and many people in the Sellwood neighborhood hope that the old bridge is refurbished for bicycle and pedestrian use. In addition, residents hope that the new bridge is safer but not significantly larger than the present Sellwood Bridge, so that traffic patterns are not drastically altered in the neighborhood.

Stephen's Creek enters the Willamette just downstream of the Sellwood Bridge. This small creek, nestled partway behind the boathouses on the west shore, could provide some key urban habitat for salmon. In 2007 the city of Portland, Willamette Riverkeeper, and other local partners worked to place woody debris in Stephen's Creek and to replant the banks, providing off-channel shade for fish in the confluence area. Although the area is small and in a congested urban setting, this streamside vegetation could make a critical difference for native fish seeking a bit of shelter from the main river.

Tides in the Willamette

Tides affect the last 26.5 miles of the river, from Willamette Falls to the confluence with the Columbia River. The tidal effects of the Pacific Ocean reach far inland, moving 100 miles up the Columbia to the Willamette and then upriver to the falls. These tides change the river level up to 3 feet or more each day. Many people who live in the Portland area don't know that the river is tidal. They might wonder why what were nice wide beaches at Sellwood Park can nearly vanish after a few hours.

9

Sellwood to Downtown Portland, RM 17 to 14

*The more clearly we can focus our attention
on the wonders and realities of the universe about us,
the less taste we shall have for destruction.*
—Rachel Carson

Sellwood Riverfront Park is on the east side of the river, just below the Sellwood Bridge. Either the dock or beach at the park, with a small adjacent parking lot, are good places to put in and take out boats in the Portland area. A new bridge in this area could well change the character of the park. Adjacent is Oaks Amusement Park, which has been in operation at the same site since 1905. These parks have been flooded numerous times over the years, with perhaps the worst occurring in the 1948 Vanport flood.

Originally this area was a peninsula that extended into the river next to a large wetland area, now Oaks Bottom Wildlife Refuge. But a railroad berm was constructed, separating the area from the river. Oaks Bottom is a fantastic urban wildlife refuge that attracts a great many birds throughout the year. Part of the property was once a dump, but today it serves as an important stopover for migratory songbirds, which find a place to rest and forage in its dense riparian fringe during the spring, as well as a host of shorebirds that can be seen traveling through. The wildlife refuge also provides an excellent base for resident birds, such as great blue heron, bufflehead (*Bucephala albeola*), and bald eagle. The bufflehead is a small diving duck that builds its nest almost exclusively in holes excavated by woodpeckers. The male has white sides, a

Stars can be seen in streaks under the Sellwood Bridge in this long exposure.

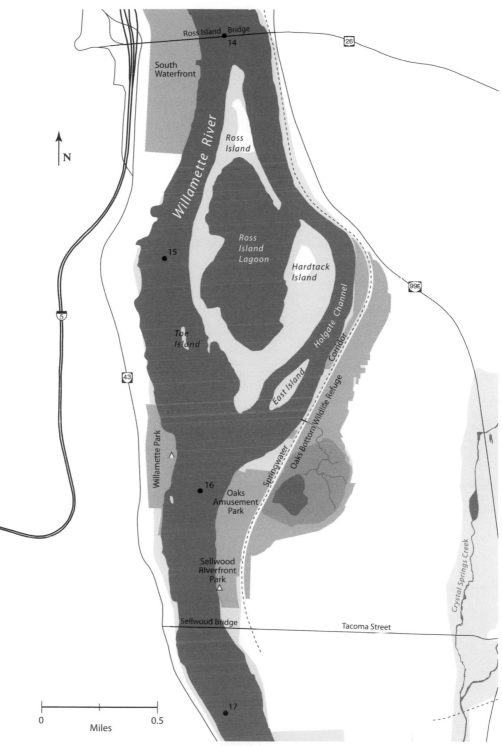

Sellwood to Downtown Portland

black back and head, and a large white head patch. Oaks Bottom is operated by the Portland Parks and Recreation Department, with significant assistance from local nonprofit groups, especially in relation to the refuge's restoration work and educational trips.

Ross Island, Hardtack Island, and East Island are located from RM 16.25 to 14. The Ross Island Sand and Gravel Company has been excavating gravel here for decades, as others did before them, and the site has long been associated with the city's growth and development. In 1926 the U.S. Army Corps of

Participants in Willamette Riverkeeper's Paddle Oregon move downriver toward the Sellwood Bridge.

The view upriver from Sellwood Riverfront Park

Engineers joined Ross Island and Hardtack Island by constructing a berm to close off the shallow dynamic channel that ran between them. This berm area has a low uniform height and is situated toward the southern end of the island on the main channel side. Rock was excavated from this cut-off channel for more than 70 years, eventually creating a large deep lagoon. Holgate Channel, which runs on the east side of the island, is typically 15 to 20 feet deep. Just past the entrance to the Ross Island lagoon, the depth drops off dramatically, quickly reaching more than 100 feet.

It is easy to imagine large Chinook canoes making their way upriver past these islands in centuries past, heading to Willamette Falls to trade with the Clackamas and Kalapuya. Perhaps these people stopped at Ross Island or Hardtack Island or maybe they just enjoyed the forested view as paddlers do today. A few minutes on a rocky and sandy beach might have been enough to stretch the legs, view the wetlands that once existed on the west side of the river, or fish in the shallows from a gravel bar.

Ross Island's lagoon is considered to be navigable water overseen by the federal government. For decades Ross Island and Hardtack Island have been owned by the Ross Island Sand and Gravel Company, except for the northern tip, which is owned by the Port of Portland. Large barges and mining equipment were a mainstay in the area, removing gravel and processing it for use in cement. The level of activity decreased slightly in 2005, however, when the

A 1958 aerial view of Ross Island and East Island. Ross Island is a much thinner shell today, after decades of gravel excavation. Note the massive collection of logs in the lagoon. Oregon Historical Society, no. OrHi55482

island's core had been dredged out and mining ceased. Today the company trucks in gravel that was mined elsewhere to be processed at the plant on Hardtack Island.

It is the responsibility of the Department of State Lands to administer public trust lands on behalf of the people of Oregon, including the bed and banks of navigable rivers like the Willamette. The department administers the permits that allowed Ross Island Sand and Gravel to dredge rock from the river. In the late 1970s, when the Ross Island company received its permit to continue dredging, a key requirement was to restore the lagoon to an average depth of 20 feet using clean fill to replace the excavated gravel. In the mid-1990s it was discovered that some contaminated fill provided by the Port of Portland had been accidentally spread around the southern end of the lagoon. At the time of its disposal this fill was placed legally, but according to standards in the 1990s the material was not considered clean fill. The Department of Environmental Quality required the company to investigate the extent and nature of the contaminated fill release. After several years of sampling and testing, the company and state concluded that the fill was not highly contaminated and the areas could be recapped with clean fill.

In 2002 when the Ross Island company acknowledged that they had mined the vast majority of the island, leaving essentially only the outer shell, they sought to rewrite their 1979 reclamation plan that was required in the previous permit. The company initiated discussions with the city of Portland, the Department of State Lands, and local environmental organizations. Together they devised a new plan in 2003 that provided for a significant amount of habitat restoration in the north and south ends of the lagoon. A total of 22 acres of wetland habitat will be restored, as well as an additional 14 acres of shallow-water habitat. In places the island will also be widened, with additional upland area created. The new and restored habitats will benefit native fish such as spring chinook, as well as shorebirds, wetland species, and upland migratory songbirds. The new plan calls for Ross Island Sand and Gravel to place far less fill than they would have had to according to the 1979 plan. Also, the monitoring and reporting plan for obtaining this clean fill is rather feeble, but there is hope that the company will place the same value on using clean fill as does the surrounding community. Despite these tradeoffs, the combination of restored habitat is far better than the original reclamation plan.

As the work to implement the new mitigation plan began in earnest in 2004, Robert Pamplin Jr., the owner of the gravel company, told then Mayor Vera Katz that he would donate the island to the city of Portland, with the exception of Hardtack Island, where the rock processing plant resides. Negotiations went back and forth for a couple of years, with a significant holdup being the work needed to address the contamination in the lagoon bottom. Once this issue was resolved in 2006, there was hope that the donation of

Ross Island could be completed. In early 2007, however, Pamplin informed Mayor Tom Potter that he was thinking of not giving the island to the city. This change of heart shifted the negotiations into high gear. Pamplin, Mayor Potter, and conservationists from Willamette Riverkeeper, Portland Audubon, and the Urban Greenspaces Institute were able to hash out an agreement that seemed to work for everyone involved. The resulting donation of 44 acres is the most undeveloped part of the island. The donated portion begins at the large bulb of land on the west side and extends north along the west shore and then inland. It does not include the shoreline of the lagoon. This agreement was formalized at a Portland City Council meeting in October 2007.

What will result from this donation of land is a bit hard to say. Conservationists hope that the 44 acres, as well as the northern portion of the island owned by the Port of Portland, with be managed as a wildlife refuge. A management plan may be initiated that incorporates Oaks Bottom Wildlife Refuge, the donated portion of Ross Island, and the part still owned by Ross Island Sand and Gravel. The agreement includes provisions for local restoration and conservation groups to access the island to help with invasive species removal and the planting of native species. Given the limited resources available and that the property can only be accessed by boat, a collaborative effort between Portland Parks and Recreation and the conservation groups is necessary to manage the property.

Like the ecological stepping stones further upriver, Ross Island can provide habitat the enables fish to rest on their way up- and downriver, as well as excellent canopy for migratory songbirds and an array of resident native wildlife. Of course, the aesthetic value of having a good-sized wildlife refuge on an urban section of the Willamette makes the project very appealing to the general public.

Access to this island will likely remain by boat only, with few options to get out of your craft. Ideas about the level of access to the island have covered a broad spectrum, from developing paths and other infrastructure for the public to allowing access only to conduct restoration activity. One key component of the restoration plan is a limitation on wakes. During the summer months many motorized craft use the lagoon for wake boarding and water-skiing, activities that stir up the interior of the lagoon to a significant degree. Such activities do not mesh well with the need to protect the habitats thus far restored, as well as future plans to protect the integrity of the island.

Across from the northern tip of Ross Island is the Zidell Company, which produces barges that are launched on a regular basis. Zidell also owns property downstream of the Ross Island Bridge, and this area has had significant contamination issues from their former business of decommissioning U.S. Navy ships. A settlement was reached with the Navy that provides some assistance in cleaning up the area, in addition to Zidell's responsibility.

Loop around Ross Island, RM 16.5 to 14

> **STARTING AND ENDING POINTS**: Sellwood Riverfront Park just off Spokane
> Avenue on the east side or Willamette Park just off Macadam Avenue
> (Highway 43) and Nebraska Street on the west side.
> **DISTANCE**: 5 miles roundtrip
> **SKILLS LEVEL**: Requires flat-water paddling experience.
> **CONDITIONS AND EQUIPMENT**: A canoe or kayak works well for this trip. Al-
> though it is all flat water, paddlers can experience wind-driven waves and
> boat wakes.
> **AMENITIES**: Both parks have parking and restrooms.
> **WHY THIS TRIP?** You can experience a large, urban restoration project and
> see wildlife, such as bald eagles, blue herons, and river otter.

This trip is excellent most of year, with the exception of high-flow periods,
which can make paddling upriver difficult. Put in at the dock or from the beach
at Sellwood Riverfront Park. You can paddle downstream along the beach, and
cross over toward Ross Island and East Island once you pass the boat houses at
Oregon Yacht Club. At low tide the area to the south of the islands becomes
a large tidal flat, replete with driftwood. In the spring, you can sometimes see
a variety of shorebirds on this flat, from spotted sandpipers to Caspian terns,
which resemble gulls but with a black-capped head and a large, pointed, coral
red bill.

Take the middle channel between East Island and Ross Island, which provides a good view of the thick riparian area. In the summer months paddlers
should be aware of large wakes from ski boats, which can rise quickly near the
shoreline. For several years in the late 1990s a sizable heron rookery was located in the middle of East Island, but as of 2007 only a few nests remained—

Heron feathers
on the beach at
Ross Island

likely because of the healthy bald eagles
nesting along the Ross Island lagoon.
Raccoons and deer have been known to
zip along the shoreline in daylight hours,
as well as the occasional beaver hunting
for tree bark.

The lagoon at Ross Island is worth a
look, given the significant amount of restoration work that is being done there.
You can travel to the interior of the lagoon with ease, but be wary of the large
tugs and barges that will continue to use
the area for the foreseeable future. Since
2000 a bald eagle nest has been located

on the western edge of the lagoon, toward the middle where the land mass is largest. If you scan the cottonwoods, the nest can be seen about a third of the way down from the top. On occasion, the eagles have been seen trying to take fish caught by the osprey in the area.

Continuing on down Holgate Channel, you can stop at the tip of Ross Island. This section of the island is owned by the Port of Portland, although the public can stop here. In the summer months this section of the island gets trashed, so take the opportunity to pick up some garbage if you can. ◼

Oaks Bottom Wildlife Refuge via the Springwater Corridor

WHERE: From Sellwood Riverfront Park to the Oregon Museum of Science and Industry (OMSI).

AMENITIES: Both access points have parking, with water and restrooms at Sellwood Riverfront Park.

Portland and the Metro Regional Government have developed a wonderful bike and pedestrian path that runs along the Willamette River from the Sellwood neighborhood to OMSI, just north of the Ross Island Bridge. The wide paved trail is designed to accommodate walkers, joggers, hikers, bicycles, wheelchairs, and strollers. The Springwater Corridor follows parts of the Springwater Railroad Line, which ran between downtown Portland and Estacada from 1903 to 1958. North of Sellwood Park, the trail winds past the 163-acre Oaks Bottom Wildlife Refuge, which is replete with wetland habitat and is a great place for birding. ◼

An early January snow blankets the dock at Sellwood Riverfront Park.

Development along the river has often been problematic, yet the South Waterfront District at RM 14.5 is putting a new face on riverside development. Located just upriver of the Ross Island Bridge is an impressive series of high-rise apartment buildings, retail, and a large facility operated by the Oregon Health & Science University. The OHSU facility, with its attached tram that ferries employees and patients up and down the hill to the main campus, serves as the anchor of this development. While South Waterfront District contains many of the traits of other urban developments, there has been a concerted effort to use green building practices by recycling and reusing some materials in the construction of these energy-efficient buildings, as well as to control stormwater runoff into the Willamette. As part of the Willamette Greenway Program, Portland Parks and Recreation and the developer are required to maintain a fringe of green along the Willamette's banks. In addition to habitat for wildlife, the area will also feature river access for residents. The hope is that many residents will use human-powered craft to make their way around the newly created Ross Island Natural Area just upriver.

Habitat restoration is a big job that requires a lot of effort, time, and money. Funding for restoration has commonly come from state and federal grant programs, such as those of the Oregon Watershed Enhancement Board, Department of Environmental Quality, Department of Agriculture, and the National Oceanic and Atmospheric Association Fisheries. Cities and towns also contribute funds to the restoration of habitat along the Willamette. In the Portland area, the Metro Regional Government has passed successful bond measures to purchase land and conduct restoration work. Other sources include private foundations and local businesses. In many cases, though, get-

Numerous houseboats were once moored in Holgate Channel, as shown in this 1920s photograph. Oregon Historical Society, no. OrHi50355

ting volunteers involved is a critical piece of any restoration project. Each year volunteers contribute thousands of hours at restoration sites within the Willamette system, as well as a good dose of moral support for this kind of work. In the coming years, it will take significant effort from federal, state, county, and municipal agencies, private foundations, nonprofit organizations, and many volunteers to implement the large-scale restoration work necessary to make the Willamette River a truly healthy waterway.

10

Downtown Portland to the Columbia, RM 14 to 0

As crude a weapon as the cave man's club,
the chemical barrage has been hurled
against the fabric of life.
—RACHEL CARSON

IN ITS EARLY DAYS, Portland was in stiff competition with Linnton, Milwaukie, and Oregon City to become the principal port along the Willamette River. The towns competed with each other for every ship that made its way slowly upriver. Portland won out over the other communities, mostly because of the depth of the river here and its civic leadership. During these early days, Portland was known as Stumptown because of the swaths of trees that had been cut down as the city developed. As more and more ships arrived, the population grew, and buildings multiplied along the river, fostering commerce within the region and beyond. Sanitation and clean drinking water became issues in the city. For a time, Portland's water was taken from the river, until the water in the area was too dirty to drink. In 1895 the city switched its water source from the Willamette to Bull Run, well outside the Willamette Watershed near Mount Hood.

The health of the Willamette River has always been an important issue for the residents of Portland. Civic leaders and residents were driven to action in the 1920s to deal with municipal wastes that fouled the water and in the 1960s to fight rampant municipal and industrial pollution. In the 1990s a renewed emphasis was placed on toxic contamination and *E. coli*, a bacteria derived from human and animal waste. Economics have always been part of the debate about keeping the river healthy. In a December 1928 editorial in the *Oregonian*, the author noted, "If home authorities do not clean up the polluted Willamette River, they may be forced to do so by the federal government.... The plan for the cleanup calls for a charge of about one-third of the normal water bill to finance the project." This early push for increased regulation regarding the cleanliness of wastewater sought to make both industry and municipalities do the right thing for the Willamette, meaning more expensive wastewater treatment. Those who generate pollution may argue that the costs of proper treatment are too high, yet the costs of inaction and the lack of

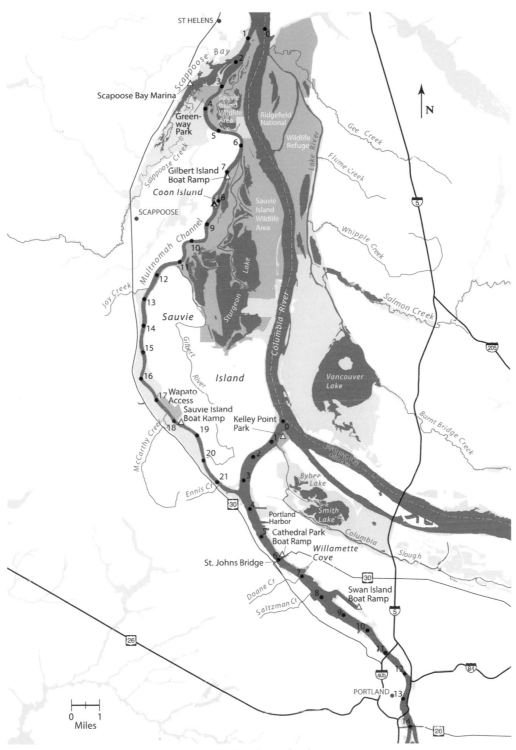

ST HELENS

Scapoose Bay

Scapoose Bay Marina

Green-
way
Park

Sauvie
Island
Wildlife
Area

Ridgefield
National

Wildlife Refuge

Gee Creek

Flume Creek

Lake River

Gilbert Island
Boat Ramp

Coon Island

SCAPPOOSE

Sauvie
Island
Wildlife
Area

Whipple Creek

Multnomah Channel

Scapoose Creek

Joy Creek

Sauvie

Sturgeon Lake

Columbia River

Salmon Creek

205

Island

Gilbert River

Vancouver
Lake

Wapato
Access

Sauvie Island
Boat Ramp

Kelley Point
Park

Burnt Bridge Creek

McCarthy Creek

Ennis Cr

30

WASHINGTON
OREGON

Bybee
Lake

Portland
Harbor

Smith
Lake

Columbia

Cathedral Park
Boat Ramp

Willamette
Cove

Slough

St. Johns Bridge

30

Doane Cr

Swan Island
Boat Ramp

5

Saltzman Cr

26

405

PORTLAND

84

0
Miles
1

26

Downtown Portland to the Columbia

thoughtful regulation and related enforcement have cost the Willamette and other rivers plenty. If a river system cannot sustain clean cold water and habitat for fish that have lived there for tens of thousands of years, then the regulations and voluntary actions are clearly not up to the challenge. In some cases, though, polluting entities can learn and change their ways. For instance, the city of Portland has modified its practices and has made a solid commitment to do more for the Willamette River, with significant progress.

Large pipes line the edge of the Willamette River in Portland. The pipe on the west side travels under the river, funneling the waste to a pump station at Swan Island, where it joins the waste from the east-side pipe. The pump station propels the sewage uphill to a large gravity-fed line that directs it to the treatment plant on Columbia Boulevard. For some time Portland's sewer system has been unable to handle the sheer volume of wastewater that is introduced to the system when it rains. Even with a small volume of rain added to the system, the sewer system overflows into the Willamette, carrying waste from homes and businesses as well as stormwater from roofs, streets, and parking lots. For many decades, the city of Portland had stalled in developing the necessary increased water treatment capacity. In 1991, however, Portland was sued because of these repeated sewer overflows, which are in clear violation of the Clean Water Act.

The result of the lawsuit was an agreement with the Oregon Department of Environmental Quality that the city would deal with its combined sewer overflow problem by 2011. Toward this aim, the city of Portland has been in-

Portland's aging sewer system must be rebuilt by 2011 to deal with the combined sewer overflow spills.

stalling massive pipes, such as the one under the river, which run parallel to the west and east sides of the Willamette, to intercept all of the sewage and stormwater that travels toward the river. This sewage upgrade is projected to cost over $1.4 billion, with much of the cost passed on to ratepayers. These massive pipes will intercept all but the largest rainstorms, and Portland will be legally allowed a handful of overflows a year. This will be a substantial improvement compared to sewage overflows almost every time it rains.

A key issue with sewage in the Willamette is what happens at the interface with people recreating along the river. *Escherichia coli* is a bacteria that lives in the lower intestines of warm-blooded animals, where it helps to properly digest food. But when *E. coli* is found in the river, it indicates fecal contamination either from sewer spills or runoff that carries animal waste to the river. *Escherichia coli* comes from various sources, ranging from human waste to that of pets, livestock, birds, and other wildlife. By itself this bacteria cannot cause illness, unless introduced into an open wound or the urinary tract. The infamous *E. coli* strain O157:H7 does create a toxic by-product that can harm humans. This strain, unlike the *E. coli* typically found in the river, is generally found in improperly prepared food. The presence of *E. coli* in the river can indicate the presence of harmful organisms such as salmonella, giardia, and others.

Portland and its residents are also making significant investments in curbing the flow of stormwater into the sewer system through green infrastructure, such as bioswales and permeable surfaces that soak in significant amounts of water. Bioswales are landscape features designed to remove silt and pollution from surface runoff. They consist of low-lying drainage courses with gently sloped sides that are filled with vegetation, compost, and/or rock. They are designed to maximize the time water spends in the swale, which aids in trapping pollutants and silt and prevents the water from running into the sewer system after a storm. Parking lots and portions of streets with permeable surfaces are seen more and more, but the vast majority of surface area is in the old-style concrete and asphalt, with little in the way to process stormwater before it heads down the grate to the sewer system. As part of its Portland Watershed Management Plan, the city has developed goals for curbing stormwater and improving habitat, along with other activities. Financial incentives and the building permit structure in the city have also encouraged innovative methods to capture stormwater.

At RM 13.5, just upriver of the Hawthorne Bridge, is the dock and building of the Portland Boathouse, the first collection of human-powered watercraft clubs and businesses in Portland. Founded in 2004, the Portland Boathouse provides storage area for several sculling teams, from youth to university and adult club levels. In addition, the retailer Alder Creek Kayak and Canoe has a shop in the building and Willamette Riverkeeper maintains its headquarters

Good river access is provided by the Portland Boathouse dock, located just upriver of the Hawthorne Bridge.

there. Establishing this facility took many years of advocacy by rowers, who were joined by others who care about river access in Portland.

The Portland Boathouse is part of the River East Center, which was formerly a large industrial warehouse called the Hollman Building. In 2004 the owners purchased it inexpensively from the Portland Development Commission in exchange for constructing a state-of-the-art, high-efficiency building with landscaping that would help to curb the flow of stormwater from their property. The River East Center directs rainwater from the roof down a common pipe that feeds into a bioswale. In addition to the Portland Boathouse, the building also houses the MacKenzie Group and Coaxis, with about 200 employees, who have helped to revitalize the area. Hopefully, the relationship between these two private companies and the Portland Boathouse tenants will continue for years to come.

The Willamette becomes more confined as it flows underneath the Hawthorne, Morrison, Burnside, Steel, Broadway, and Fremont Bridges. The bank is relatively hardened in many places, with the notable exception of the Vera Katz Esplanade, which stretches along the east bank of the river between the Hawthorne and Steel Bridges. Interstate 5 is just feet away from the east riverbank along this stretch, but the esplanade provides an opportunity for many walkers, runners, and bicyclists to see the river up close. Some native plantings along the esplanade have helped to add a little bit of nature to this busy area, with wildlife in the area taking note as well. In 2007 beavers made their way up the bank and tried to chew the bark off the newly planted trees, as they do elsewhere along the river. Portland, which is known for its progressive views toward wildlife, worked to make the job a bit tougher by erecting chicken wire barriers around the trees on the esplanade. The city chose not to remove the beavers, however, because the animals would likely just come back as long as the food remained.

On the west riverbank in front of the high-rise buildings of downtown, the defining feature is a 20-foot-tall concrete seawall that divides downtown from the river. Portland has suffered numerous major floods—in 1864, 1964, and 1996—and the city has seen its share of river water covering downtown streets. As a result, the idea of having a hardened structure to keep the river at bay has gained acceptance. Yet throughout most of Tom McCall Waterfront Park the river is 20 feet below the walkway even though the park stretches from the Interstate 5 bridge to the Steel Bridge in downtown Portland. There is little opportunity to get to the water's edge to actually splash along the shore or touch the water, which makes getting a canoe or kayak into the river very difficult.

Over a 6-mile stretch in the heart of the city, from the Sellwood Bridge to the Fremont Bridge, there are essentially two good access points on the east side (Sellwood Riverfront Park and the Portland Boathouse dock) and three good access points on the west side (Willamette Park, the large sloped lawn just upstream of the Hawthorne Bridge, and Riverplace Marina). Near Oaks Bottom Wildlife Refuge, the Springwater Corridor trail runs a fine course parallel to the river's east bank, small footpaths to the river's edge offering some points of access and excellent views of Ross Island. However, most of the "access" to the river on this side is from relatively steep banks or privately owned land. On the bright side, the city has begun to create more access points to the river, and the Portland Boathouse dock on the east side is a good example of this. Additional access points are planned, such as a public dock in the South Waterfront District.

Of all the river's miles, this stretch along Portland has been modified the most. From the city's earliest days, commerce and industry used what the Willamette River had to offer. The city's port welcomed ships and riverboats from near and far, facilitating the transport of all manner of goods. Factories for steel production, wood treatment, cement fabrication, ship building, petroleum processing and storage, and other industries lined the Willamette as well. Portland has always had a heavy industrial footprint along the river, particularly along Portland Harbor, which stretches from the Steel Bridge at RM 12 to the last hardened surfaces where the Columbia Slough enters the Willamette at RM 1. Portland Harbor has multiple embayments or slips that allow ocean-going ships to be backed into their terminals to load and unload. At nearly every spot along the river's shoreline, the bank is hardened with rock,

A very unnatural bank in Portland Harbor

metal, pilings new and old, dock structures, cement walls, and other debris. Only a few points have somewhat natural settings, with a sloped beach and a backdrop of trees. The more than 100 years of industrial activity in Portland Harbor have had dramatic negative impacts on the health of the Willamette in this stretch.

From around the Broadway Bridge to the confluence with the Columbia, the U.S. Army Corps of Engineers maintains the Willamette at a uniform depth of 40 feet. Ships fully loaded with grain routinely require these depths to pass safely through the channel, and the dredging of riverside areas near the docking slips is needed to allow loading and unloading. Consequently, the river bottom in this area is unnatural, a giant 40-foot-deep U shape in the main channel, with relatively few shallow areas near the river's edge.

Since 1995 there has been significant debate about the need to dredge the Columbia River to a depth of 43 feet, thereby allowing newer container ships with deeper drafts and a much larger overall size to make it upriver to Terminal 6. Some have also advocated a deeper Willamette in the Portland Harbor area. There are questions, however, about the future of shipping in the Lower Willamette River. The location of the Port of Portland 100 miles from the Pacific Ocean is a geographic hurdle that may cast doubt on the port's potential for container shipping in future decades. Another issue affecting the dredging is that RM 8 to 2 is at the heart of the Willamette River's Superfund site.

In May 1998 the Oregon Department of Environmental Quality issued the "Portland Harbor Sediment Investigation Report" about the breadth of contamination in the area, with its nature and extent mapped out to some degree. Various pollutants were found on the river bottom, mixed in the rich brown mush of river sediment. The vast majority of the area was built out long ago, with a variety of dock structures, concrete walls, and riprap. Many of these upland riverside areas remained contaminated, despite the cleanup efforts of earlier decades, and in some cases groundwater still fed contaminants to the river. This 6-mile stretch of Portland Harbor was a huge polluted mess, with toxic compounds such as DDT, PCBs, and PAH (derived from petroleum products), as well as heavy metals and other pollutants scattered throughout the area.

The state-led effort for a voluntary cleanup fizzled out, leading to a federal designation of the area under the Comprehensive Environmental Remediation Cleanup Liability Act (CERCLA), the federal law most often referred to as "the Superfund," in December 2000. The name Superfund has led to a general misconception that vast funds are automatically triggered to clean up a polluted site. Originally, CERCLA provided for a tax on the petroleum and chemical industries to generate funds that could be used when polluting companies fought their cleanup obligations. Instead of waiting what could be years for the Environmental Protection Agency (EPA) and a polluter to agree on a course of action, the EPA could use funds from the account to take action in the near term. If a company fought action, it could be fined three times the amount of the cleanup cost. In 1995 this provision of the Superfund ran out and the tax was discontinued. Today Congress appropriates monies, usually too little to address the scope of cleaning the many Superfund sites across the country. In addition, without the threat of large fines to the polluters if they fight cleanup, the EPA has far less leverage than it had previously.

In the case of Portland Harbor, the parties involved in polluting the area— ironically known as potentially responsible parties (PRPs)—include private companies as well as public entities. Initially, this process benefited from a good amount of cooperation among the EPA and the various PRPs. In the ideal scenario, these entities would work together to meet their combined responsibility to clean up their mess. As part of the remedial investigation to decide who is responsible for which contaminants, significant research is conducted on the types of pollution in the river and their likely sources. Following this, a feasibility study is performed to examine likely remedies for the contaminated river sediment and upland areas. After this, CERCLA prescribes that a record of decision be developed, providing the final cleanup plan for a given site. In a simple case with one company, some level of back and forth might occur, but the EPA makes the final decision and requires that company take the level of action necessary to get the job done. The case of Portland Harbor is extremely complex, however, as more than seventy PRPs have been identified. All of the PRPs are either current landowners or tied to the land legally. According to CERCLA, if someone buys contaminated land, they are responsible for its cleanup.

The EPA must also consult with numerous trustees, which include federal natural resource agencies such as the National Oceanic and Atmospheric Association Fisheries as well as local Native tribes. The tribal trustees at Portland Harbor are the Confederated Tribes of the Warm Springs Reservation of Oregon, Confederated Tribes of the Umatilla Indian Reservation, Confederated Tribes of the Grand Ronde Community of Oregon, Confederated Tribes of Siletz Indians of Oregon, Nez Perce Tribe, and Confederated Tribes and Bands of the Yakama Nation. The stretch of the Willamette in question is part of the historic fishing and hunting grounds of these tribes. Their "usual and accustomed" places were protected in the Treaty of 1855, which maintained Native peoples' right to continue fishing, hunting, and gathering from the land and water and provided partial compensation for the millions of acres the tribes ceded to the U.S. government.

Mitigating damage to natural resources is another part of the Superfund cleanup process. Tribal representatives and state and federal natural resource agencies, such as the U.S. Fish and Wildlife Service, work with the EPA and PRPs to determine the level of habitat restoration work that should occur. The damage assessment only takes into account damage to water, air, land, and biota after 1980, when CERCLA was signed into law. A formal plan for restoration must be developed and approved by the trustees and could call for the PRPs to fund a range of projects, from riparian restoration to land acquisition. Along this section of the Willamette some properties that are not highly contaminated might eventually be purchased for habitat restoration. A great deal has been invested in the habitats far upstream in the Willamette system. If the cleanup of Portland Harbor is not comprehensive and timely, then the impact of those upstream efforts will be diluted. Anadromous fish, such as spring chinook, must pass through Portland Harbor before they ever have a chance of taking advantage of healthy habitat upstream and access to their natal streams. Perhaps in a few decades, this area too will have clean water, clean sediment, decontaminated upland areas, and restored habitat.

As you make your way downriver, take a close look at the west side. Imagine some 150 years ago, back before any of the large buildings, cranes, and massive ships were here. Instead, you would have seen a vast connected wetland area reaching toward the hills of what is today Forest Park, with the calls of geese and other waterfowl punctuating the sounds of the flowing river. Part of this area, known originally as Guilds Lake, was a 400-acre wetland complex situated just north of the Fremont Bridge. The southern portion of the site was used in 1905 for the Lewis and Clark Exposition, a large fair that celebrated Portland and the region. Most of the structures for the exposition were not built to last and were left to decay or were replaced by industrial development as the wetland was filled. By the 1920s most of this wetland was gone. As you travel the Willamette River today, it can be difficult to envision what

was a vast tidal waterway, unencumbered by the tight control of dams high in the river's tributaries. Large flat islands of rock and sand and substantial sandy shoals were the order of the day.

Swan Island was once a large tidal island, yet today it is a highly industrialized part of Portland Harbor. In 1930 Swan Island was joined to the east bank of the river, making it a peninsula. At RM 10 a beach and riverside area of uniform height marks the place where the main channel of the Willamette once ran through this area. As you travel past Swan Island, you can see ships moored or in drydock being repaired and maintained. This portion of the island has long been affiliated with the construction and repair of ships. During World War II, Liberty ships were built here, originally with Keizer Steel constructing the ships for the U.S. Navy. When the largest drydock on the west coast left Portland in 2003, the main marine contractor, Cascade General (now Vigor Industrial), lost some significant repair capacity. Many ships are still repaired and maintained in this Swan Island facility, but not the largest ships. Today Vigor Industrial builds many barges at this site, continuing this maritime tradition.

Overlooking the Swan Island area is a large bluff at RM 8 where the Univer-

A 1927 photograph of the Portland Harbor area of the river. The large area in the upper left of the photograph was part of Guilds Lake, which was soon filled in and built upon. Also note Swan Island on the right, still covered in trees. Oregon Historical Society, no. OrHi49925

sity of Portland is located. Although the university has overlooked the river for some years, they have never owned riverfront land. At present they have a keen interest in an area known as Triangle Park, a former industrial site. The university envisions adding to its sports complex at this site. Although practice fields are not exactly a nature preserve, this use would be quite new in the harbor area.

William Clark ascended to this point on the river back in 1805. The Lewis and Clark party had missed the Willamette River as they traveled along the Columbia River's north shore, which is understandable given the highly braided channel where the Willamette entered the Columbia, with much of the flow heading down Multnomah Channel at that time and with a series of islands obscuring their view. After being informed by the Natives downriver that they had passed the Willamette, Clark headed back and witnessed Chinook villages on Sauvie Island. It is said he ascended the riverside to a broad open area, affording a view upriver. Many have agreed that the general area of the bluff overlooking Swan Island is where he ended up, though some over the years have made a case that Clark made it further upriver.

In his journal, Clark stated, "I entered this river which the natives had informed us of, called Mult no mah River so called by the natives from a Nation who reside on Wappato Island a little below the enterance of this river. . . . I

The old railroad bridge crosses the Willamette downstream of Swan Island.

proceeded up this river 10 miles from its enterance into the Columbia to a large house on the NE Side and Encamped near the house. . . . at this place I think the width of the river may be Stated at 500 yards and Sufficiently deep for a Man of War or Ship of any burthen." If only Clark could have seen what the river would look like 200 years later. Although a large part of the Lewis and Clark expedition had to do with commerce and positioning the young United States with the world powers of the day, surely Clark never would have imagined the level of pollution that now exists in this portion of the river.

Just downstream of Swan Island adjacent to the railroad bridge is the McCormick and Baxter property, which was an earlier Superfund site that had its major cleanup activity finalized by February 2005. This property was highly contaminated by creosote, which was widespread in the groundwater table and eventually contaminated sediments on the river bottom and 50 feet below. The main cleanup activity removed a mass of soil from the upland area, and then a sheet pile wall was constructed to keep the remaining contaminated groundwater from continuing to make its way toward the river. The underwater portion of the contaminated area could not be dredged, due to the risk of polluting the river downstream, so the area was capped with clean fill. Because of the size and level of pollution on the McCormick and Baxter property, long-term monitoring will be required to ensure that new issues do not arise with the creosote that still remains underground. This is a problem for all contaminated sites that are capped.

In some ways, areas like Portland Harbor are easy to rationalize. The industries here, after all, created and still do create many good jobs. They promote international trade and help stimulate the economy of Oregon and the region. The area also must be viewed in its historical context. Much of the development in Portland Harbor goes back more than 100 years, and our society's values related to keeping our environment clean have broadened and become stronger over the decades. But ample opportunities remain to treat the river better. The challenge of Portland Harbor and other industrial areas can be met.

Even in riverside areas that have significant industrial use, riverbanks can be modified to better accommodate wildlife and provide some level of healthy habitat for fish. The traditional hardening of banks could be decreased, and the loading and unloading of ships could occur with docking structures placed further out from the bank. Certainly stormwater can be better filtered in many cases as well. Even in relatively small lots, natural areas interspersed among the industrial sites could provide valuable habitat for fish and other wildlife. If you look at Portland Harbor from the air, you can see that the numerous creeks running off the west hills become enclosed as they enter the bottomlands. Perhaps more of these historic connections can be opened up and the creeks' confluence areas improved. Yet for all of this to occur, it will

Cormorants perched on pilings in Portland Harbor.

Double-crested Cormorant

Double-crested cormorants (*Phalacrocorax auritus*) can be seen upriver at times, but these large black water birds gather in greater concentrations in the Portland Harbor area, as well as the Lower Columbia. Cormorants frequently spread their wings out to dry them while perched on pilings and wires, though they have a layer of waterproof feathers next to their skin. Cormorants search out fish by diving under the surface of the river and propelling themselves along with their webbed feet. It is not usual to hear them call out, but if you are able to get close to their roosts, you can hear a series of grunts and wails. If you encounter some double-crested cormorants sitting together on a long wire perched over the water, they will typically fly off quickly if you get too close, so give the birds a little space.

take more than the existing regulation, incentives, and sources of funding. It will take a philosophical change in our society to make Portland Harbor and other areas of the Willamette healthy enough to provide the natural processes so vital to fish, birds, mammals, and even people.

Aldo Leopold, the legendary conservationist, noted that society had not yet reached the point where we understand it is in our interest to protect a local stream from contamination, a mountainside from clearcutting, or the air from being despoiled by pollution. A basic premise in Leopold's essay "The Land Ethic" (Leopold 1949) is that people too often rely on the government to do something and fail to do it themselves, unless there is some economic gain to be had. This is one of the chief issues in conservation and restoration today: myriad incentives exist to help landowners do the right thing, yet many argue for more. In *The Round River* (1953), Leopold said, "In our attempts to save the bigger cogs and wheels, we are still pretty naïve. A little repentance just before a species goes over the brink is enough to make us feel virtuous. When the species is gone we have a good cry and repeat the performance." Although there are laws in place to protect the public trust, what it really comes down to is the willingness of people to do the right thing on their own. Certainly the opportunity is there, whether looking at the floodplain lands upriver of Harrisburg or the streams that descend the flanks of Portland's Forest Park.

If Portland Harbor or the diminished habitats far upstream are any indication, we have fallen well short of understanding the need to do the right thing at a personal level. As in the cases of spring chinook, Oregon chub, and numerous other threatened and endangered species, our society seems to do just enough to show that we are doing more than nothing. Unfortunately, this is a far cry from providing the scale of action on the ground to get the job done.

Beavers still make a living in Portland Harbor, as shown by these sticks stripped of their bark at Cathedral Park.

However, increased levels of voluntary action, coupled with public and private investment, may indicate that more people are "getting it." Hopefully, in time our understanding and appreciation for the Willamette will help foster a new river ethic.

Just beyond the railroad bridge at RM 7 is Willamette Cove, which was purchased by the Metro Regional Government with funds from a 1995 bond measure. With some relatively minor contamination issues, the site holds promise for some form of habitat restoration work to occur. Sites like this are few and far between in the Portland Harbor area, although there is hope that additional green-

A tank farm can be seen through the pilings in Portland Harbor.

way areas can be established even in this heavily industrialized section of the Willamette.

The St. Johns Bridge, at RM 6, lies across a wide open section of the river in the heart of Portland Harbor. The bridge is home to nesting pairs of peregrine falcons that have successfully fledged chicks. The high vantage point mimics the native cliffside habitats of these birds, enabling them to easily seek prey, such as pigeons. Below the bridge in Cathedral Park there is good river access from a boat ramp and small dock. In the spring months the area can be extremely busy, as many people who fish for spring chinook use this boat ramp. It is also a prime area for sturgeon fishing, which also underscores the importance of ensuring that the toxic contamination in this area of the river is cleaned up.

Cathedral Park

WHERE: Cathedral Park can be accessed from North Edison Street and Pittsburgh Avenue at the east side of the St. Johns Bridge.
AMENITIES: Restrooms and free parking.

Cathedral Park is the site of one of Portland's earliest ferries, operated by James John in the 1850s. Next to the boat ramp and dock you can see upriver and down, getting a good view of the built-up riverbanks and other development, such as the tank farm just across the river that stores petroleum products. Cathedral Park is a good vantage point to see the prevailing industrial use of the Portland Harbor area. ∎

Just past the St. Johns Bridge on the east side of the river is the Toyota Terminal, part of the Port of Portland's Terminal 4. Autos and trucks are offloaded and customized at this finishing facility, before being sent by truck and train throughout the western United States. The Toyota facility is a good neighbor to the river, in that it has worked to curb stormwater runoff from the vast parking lot and has created a riparian buffer that provides some habitat value. The facility is certainly more river-friendly than others along this stretch of the Willamette.

As you make your way through this stretch, there is little in the way of natural space or green. Here and there are a sandy beach or two, with pilings from old docks and other riverside structures dotting the river's edge. But if traveling in a small paddle craft, you can slip behind the docks and pilings and occasionally be greeted by Canadian geese and other birds along the shoreline.

Cathedral Park to Kelley Point Park, RM 6 to 0

STARTING AND ENDING POINT: Cathedral Park can be accessed from North Edison Street and Pittsburgh Avenue at the east side of the St. Johns Bridge.
DISTANCE: 12 miles roundtrip
SKILL LEVEL: Experience with wind-driven waves and large wakes.
CONDITIONS AND EQUIPMENT: Works well for sea kayaks and canoes. Large tugs and boats can create significant wakes, which can be especially difficult if you are next to hardened surfaces where the waves rebound.
AMENITIES: Parking, a boat ramp and dock, and restrooms.
WHY THIS TRIP? You can see a large section of river that is highly industrialized and is undergoing a significant cleanup. The wildlife here may surprise you.

Paddlers must be very wary of large tug and ship traffic in this part of the river. The wakes can be large, but they are manageable. Also, if you cross the channel here in a paddle craft, make sure to continue looking up- and downstream. This part of the river is wide, and boats may seem further away than they actually are.

From Cathedral Park you can make your way downriver about 6 miles to Kelley Point Park. Just shy of the Columbia River, you will see the Columbia Slough flowing into the Willamette on river right. Along this stretch most of the shoreline is built up, but in a paddle craft, you can slide along the inside of some of the dock structures and piers, edging along that thin strip of water next to the shoreline. Here you may get a closer glimpse of double-crested cormorants, ospreys, bald eagles, and great blue herons.

By heading upriver about 1 mile along the riprapped bank from Cathedral Park, you can access Willamette Cove, an inviting spot on this section of river. The cove has an abundance of black cottonwoods and other trees in the riparian area, and the beach is gradual and sandy. Unfortunately Willamette Cove has some contamination issues to be dealt with, and McCormack and Baxter next door is a treated Superfund site that has ongoing monitoring. Be sure to stay in public areas in this stretch of river. Just upstream of Cathedral Park is the city of Portland's laboratory, where numerous water-quality tests are conducted. Walk the grounds if you have the time, as many features have been built to address the flow of stormwater from the property. ▣

The southern tip of Sauvie Island is at RM 3, where the 21-mile Multnomah Channel meets the Willamette. The Columbia runs along the eastern shore of Sauvie Island. This large island was once a key area to the Chinook people, who had several town sites with long houses here. The Chinook thrived on Sauvie Island, with its relatively mild climate and abundance of fish, wapato, and game. One large village was centered on the southeastern tip of the island, not far from where Multnomah Channel begins. Today, Multnomah Channel is about a third of the width of the mainstem Willamette in this stretch. For much of the next 21 miles a uniformly constructed dike separates the channel from the island, and the channel is hemmed in on the west side as well. Much of Multnomah Channel is punctuated with houseboats and moorages. Traveling this channel can be somewhat monotonous, though there are areas of interest, especially near the Gilbert Island boat ramp about 10 miles up the channel, with access during part of the year to the Sauvie Island Wildlife Area.

The 12,000-acre Sauvie Island Wildlife Area is managed by the Oregon Department of Fish and Wildlife. The area is open to the public only from mid-April through September. About 250 species of birds can be seen on the island throughout the year. In autumn and winter, it hosts more than 150,000 migratory ducks and geese. Other seasonal visitors include bald eagles in winter, sandhill cranes (*Grus canadensis*) in autumn and spring, and tundra swans (*Cygnus columbianus*) in autumn. Great blue heron, wood ducks, red foxes, raccoons, beavers, and black-tailed deer live on Sauvie Island throughout the year. The wildlife area is a good place to hike, and be sure to bring some binoculars. When they are open to the public, you can also paddle into Gilbert River and Sturgeon Lake in the wildlife area.

At 2 miles from the end of Multnomah Channel is the entrance to Scappoose Bay, which provides an interesting look at tidal habitat. At low tide, it turns to an expanse of mud flats broken by small channels of water that may allow passage for a small boat. Situated between Scappoose and St. Helens, the bay has a boat ramp and moorage, as well as several houseboats. The west shore of Scappoose Bay was once home to a Chinook village, in what is today a gently sloping grassy expanse where cattle graze. This area has been excavated numerous times and yielded interesting clues about the people who lived here and the life they led. This site had twenty-eight houses and more than 1000 residents. Thousands of arrow heads and other relics such as figurines, paint pots, and mauls were found here.

This village was indicative of life in the area long before the appearance of Lewis and Clark and is similar to other villages around Sauvie Island and elsewhere along the Lower Columbia and Willamette Rivers. Here it is easy to imagine the long dugout canoes pulled up along the gradual incline of the shoreline, with multiple longhouses at the top of the rise. The Chinook village had access to the open water of the Columbia River only a few miles away, yet the bay sheltered it from the brunt of flood events. The nearby hills and local drainages, such as Scappoose Creek, provided access to the Tualatin Valley and points beyond. The natural resources in the area, including sturgeon, salmon, and wapato, were good staples to carry them through the seasons. In his book *Naked Against the Rain*, Rick Rubin gives a wonderful description of this culture and the land that sustained it.

The Columbia Slough enters the Willamette at RM 1. For many years the slough was highly polluted, with all manner of trash dumped in it. Since the early 1990s, however, the city of Portland and Columbia Slough Watershed Council have made good strides in cleaning up the slough and conducting restoration work. Although the cleanup has a long way to go, there is good momentum for making additional improvements. A variety of wildlife can be seen all along the Columbia Slough, with the bird activity being surprising at times.

Between the slough and the Columbia are Smith and Bybee Lakes, a wonderful refuge area in the heart of the North Portland industrial area. At nearly 2000 acres, Smith and Bybee Lakes Wildlife Area is the largest protected urban wetland in the United States. The area can be experienced by paddle craft or on foot via the paved Interlakes Trail, which can be accessed at the parking area on Marine Drive. The trail is less than a mile long and has two wildlife-viewing platforms. Smith Lake is maintained at a fairly constant water level and has been developed for low-intensity water recreation, such as fishing and canoeing, whereas Bybee Lake is allowed to rise and fall with the tides and seasons and is kept as an environmental preserve. More than 100 species of birds have been recorded in the area, including great blue herons, red-tailed hawks, and several species of waterfowl. Ospreys nest at Smith and Bybee Lakes, and bald eagles winter here. Many migrating songbirds and shorebirds stop at the wildlife area as well.

Kelley Point Park provides a ready place to stroll the beach and take in the wide open confluence of the Willamette and Columbia Rivers. The Willamette does not end with a rush. Instead the vast greenish blue pulse of water enters the larger Columbia in a broad expanse, serving as a gateway to ships, tugs, sailboats, canoes, kayaks, and other craft.

Kelley Point Park

WHERE: From Interstate 5 take Exit 307, turn right onto North Marine Drive, and follow it for 4.5 miles to North Kelley Point Park Road.
AMENITIES: Restrooms and free parking.

Kelley Point Park is located on the far northwest tip of the peninsula of Portland. This 104-acre park has paved and unpaved paths, as well as a large sandy beach. Kelley Point attracts residents for its shoreline fishing opportunities, dog walking, and river viewing. Here you can witness the last mile of the Willamette and see the open expanse of the Columbia River beyond. ■

For hundreds of miles the collected waters that flow into the Willamette end up here. Snowmelt that feeds Olallie Spring and the Middle Fork in the Cascades meshes with water from the Coast Fork fed from the Coast Range. The McKenzie, Long Tom, Marys, Santiam, Luckiamute, Yamhill, Pudding, Mollala, Tualatin, and Clackamas Rivers and hundreds of creeks join the flow as the Willamette heads to the Columbia. From the vantage point on the beach at Kelley Point, you are witness to a river's end. Yet the Willamette never ends, as the water continuously percolates from the high snow fields and up through the basalt to join the vast Willamette system. The Willamette River is linked to the fate and function of so many rivers in the Willamette Basin, just as the Columbia is linked to the fate of the Willamette River. In many ways, each of us is responsible for the fate of these great river systems. To achieve and maintain a healthy Willamette River, we must:

▸ Call upon federal, state, county, and municipal agencies and others to direct funding to restoration projects on the Willamette River and its tributaries
▸ Adequately fund the implementation of the Clean Water Act in Oregon and elect leaders who will wisely enforce this important law
▸ Comprehensively clean up contaminated sites, such as Portland Harbor
▸ Stay on watch for contamination with good monitoring and sampling
▸ Financially support and volunteer with those organizations working for habitat restoration
▸ Curb runoff from the myriad nonpoint sources, such as roofs, lawns, parking lots, streets, and agriculture
▸ Keep contaminants out of the water in your yard, driveway, and local creek
▸ Develop a new river ethic by getting on the river to experience and learn about it, then engage in protecting it

In coming years there is a very real opportunity to improve the condition of the Willamette River, from the cleanliness of its water to the health and abundance of its habitat. This road will be a long and difficult one, but we owe it to the river and to ourselves to make this effort. Each of us can be part of the story and part of the solution. By learning about how the Willamette River looks and feels and by experiencing it—if only through the pages of this book—you have become part of the solution. It is now your turn to share with others what you have learned about this great river. The Willamette has so much to offer, and the responsibility rests on us to take care of this natural wonder for now and for generations to come.

Resources

River and Riparian Conservation and Restoration

Willamette Riverkeeper
1515 SE Water Ave., #102
Portland, OR 97214
(503) 223-6418
www.willamette-riverkeeper.org

Tualatin Riverkeepers
12360 SW Main St., #100
Tigard, OR 97223
(503) 620-7507
www.tualatinriverkeepers.org

McKenzie River Trust
1245 Pearl St.
Eugene, OR 97401
(541) 345-2799
www.mckenzieriver.org

Oregon Clean Water Action Project
212 Pearl St., # 1
Eugene, OR 97401
(541) 686-3027
www.oregoncleanwater.org

Friends of Smith and Bybee Lakes
PO Box 83862
Portland, OR 97283-0862
www.smithandbybeelakes.org

Friends of Buford Park and
    Mount Pisgah
PO Box 5266
Eugene, OR 97405
www.bufordpark.org

West Eugene Wetlands
751 S. Danebo Ave.
Eugene, OR 97402
(541) 683-6494
www.wewetlands.org

Eugene Stream Team
1820 Roosevelt Blvd.
Eugene, OR 97402
(541) 682-4850
See link at www.eugene-or.gov

Oregon Environmental Council
222 NW Davis St., #309
Portland OR 97209-3900
(503) 222-1963
www.oeconline.org

Urban Greenspaces Institute
Department of Geography
Center for Spatial Analysis and
    Research
Portland State University
Cramer Hall, Room 459
Portland, OR 97228-6903
(503) 319-7155
www.urbangreenspaces.org

Greenbelt Land Trust
PO Box 1721
Corvallis, OR 97339
(541) 752-9609
www.greenbeltlandtrust.org

Audubon Society of Portland
5151 NW Cornell Rd.
Portland, OR 97210
(503) 292-6855
www.audubonportland.org

Oregon Network of Watershed
   Councils
PO Box 40061
Eugene, OR 97404
www.oregonwatersheds.org

Watershed Councils along the Willamette

McKenzie River Watershed Council
PO Box 70166
Eugene, OR 97401
(541) 687-9076
www.mckenziewc.org

Marys River Watershed Council
311 SW Jefferson Ave.
Corvallis, OR 97333
(541) 758-7597
www.mrwc.net

Luckiamute River Watershed Council
Western Oregon University
345 N Monmouth Ave.
Monmouth, OR 97361
(503) 838-8804
http://luckiamute.watershed
councils.net/

Johnson Creek Watershed Council
1900 SE Milport Rd.
Milwaukie, OR 97222
(503) 652-7477
www.jcwc.org

Columbia Slough Watershed Council
7040 NE 47th Ave.
Portland, OR 97218-1212
(503) 281-1132
www.columbiaslough.org

Government Agencies

Oregon State Marine Board
PO Box 14145
435 Commercial Street NE, #400
Salem, OR 97309-5065
(503) 378-8587
www.boatoregon.com
Enforces marine law, provides boating
safety education and boat registration

Oregon Department of Environmental
   Quality
811 SW 6th Ave.
Portland, OR 97204-1390
(800) 452-4011
www.oregon.gov/DEQ
Aims to maintain water quality and
issues related permits under the Clean
Water Act

Oregon Department of Fish and
   Wildlife
3406 Cherry Ave. NE
Salem, OR 97303-4924
(800) 720-6339
www.dfw.state.or.us
Helps to protect native wildlife, and
consults with other agencies on
impacts to the state's species

Oregon Department of State Lands
775 Summer St. NE, #100
Salem, OR 97301-1279
(503) 986-5200
www.oregon.gov/DSL
Regulates removal and fill permits that can affect wetlands, in addition to properties along the Willamette that could be dredged; consults with the U.S. Army Corps of Engineers

Oregon Department of Agriculture
635 Capitol St. NE
Salem, OR 97301-2532
(503) 986-4550
www.oregon.gov/ODA
Regulates agricultural practices relating to runoff from agricultural operations. Most of this agency's work is complaint driven. In recent years, they have sought to increase riparian restoration on private lands.

Oregon Parks and Recreation
    Department
725 Summer St. NE, #C
Salem, OR 97301
(800) 551-6949
www.oregon.gov/OPRD
Manages many of the greenway sites along the Willamette River

National Oceanic and Atmospheric
    Association Fisheries
(301) 713-2334 Ext. 174
www.nmfs.noaa.gov
The federal resource agency that implements the Endangered Species Act relating to the Willamette's spring chinook and other threatened species that are anadromous (ocean going). This

agency theoretically has the power to make other agencies do the right thing for the river.

U.S. Fish and Wildlife Service
Oregon Office
2600 S.E. 98th Ave., #100
Portland, OR 97266
(503) 231-6179
www.fws.gov
Holds authority relating to the Endangered Species Act for species that are not anadromous; operates the Federal Wildlife Refuges in the Willamette Valley

Metro Regional Government
600 NE Grand Ave.
Portland, OR 97232-2736
(503) 797-1700
www.metro-region.org
Maintains parks and nature trails in the Portland metropolitan area

Portland Bureau of Environmental
    Services
1120 SW 5th Ave., #1000
Portland, OR 97204-3713
(503) 823-7740
www.portlandonline.com/bes
Provides water-quality protection, wastewater collection and treatment, and sewer installation

Portland Parks and Recreation
1120 SW Fifth Ave., #1302
Portland, OR 97204
(503) 823-7529
www.portlandonline.com/parks

Other Resources

Willamette Basin Explorer
www.willametteexplorer.info

Northwest Steelheaders
PO Box 22065
Milwaukie, OR 97269
(503) 653-4176
www.nwsteelheaders.org

Trout Unlimited
www.tuoregon.org

Water Watch
213 SW Ash St., #208
Portland, OR 97204
(503) 295-4039
www.waterwatch.org

River Safety and Navigation Resources

American Canoe Association
7432 Alban Station Blvd., #B-232
Springfield, VA 22150
(703) 451-0141
www.acanet.org

Lower Columbia River Canoe Club
www.l-ccc.org

Willamette Kayak and Canoe Club
PO Box 1062
Corvallis, OR 97339
www.wkcc.org

Cascade Canoe Club
c/o Horst Lueck
260 E 2nd Ave. #6
Eugene, OR 97401
(541) 687-5796
http://canoe.freeshell.org/

Willamette Falls Locks
U.S. Army Corps of Engineers
West Linn, OR 97068-3397
(503) 656-3381
www.us-parks.com/usace/
willamette_falls_locks

U.S. Geological Survey, for information
on river levels
http://waterdata.usgs.gov/or/nwis/rt

Westfly, for information on river levels
http://www.westfly.com/cgi-bin/
riverData?region=OR

Bibliography

Allen, Jennifer, Autumn Salamack, and Peter Schoonmaker. 1999. Restoring the Willamette Basin: Issues and Challenges. Salem, Ore.: Willamette Restoration Initiative.

Benner, Patricia. 2005. *The Willamette River near Corvallis: River History and Ecology*. Philomath, Ore.: Benton County Historical Society.

Campbell, Ron, and Jack Wiles. 2006. Luckiamute State Natural Area Master Plan Draft. Salem, Ore.: Oregon Parks and Recreation Department.

Carson, Rachel. 1962. *Silent Spring*. Boston: Houghton-Mifflin.

Cornell Lab of Ornithology. All about birds. Available via www.birds.cornell.edu/AllAboutBirds. Accessed April 2008.

Corning, Howard McKinley. 1947. *Willamette Landings: Ghost Towns of the River*. Portland, Ore.: Binfords & Mort.

Cronin, John, and Robert Kennedy Jr. 1997. *The Riverkeepers*. New York: Scribners.

Curtis, Larry, Kim Anderson, and Jeffrey Jenkins. 2004. Environmental Stresses and Skeletal Deformities in Fish from the Willamette River. Corvallis: Oregon State University, Department of Environmental and Molecular Toxicology.

Elder, Don, Gayle Killam, and Paul Koberstein. 1999. *The Clean Water Act: An Owner's Manual*. Portland, Ore.: River Network.

Elphic, Chris. 2001. *The Sibley Guide to Bird Life and Behavior*. New York: Alfred A. Knopf.

Gabrielson, Ira N., and Stanley G. Jewett. 1940. *Birds of Oregon*. Corvallis: Oregon State College.

Giordano, Pete, for the Willamette Kayak and Canoe Club. 2004. *Soggy Sneakers*. Seattle, Wash.: Mountaineers Books.

Gordon, Steve, and Cary Kerst. 2005. *Dragonflies and Damselflies of the Willamette Valley, Oregon*. Eugene, Ore.: Crane Dance Publications.

Hulse, David, Stan Gregory, and Joan Baker. 2002. *Willamette River Basin Planning Atlas: Trajectories of Environmental and Ecological Change*. Corvallis: Oregon State University Press.

Hussey, J. A. 1967. *Champoeg: Place of Transition*. Portland: Oregon Historical Society.

Jones, Philip N. 1997. *Canoe and Kayak Routes of Northwest Oregon*. 2nd ed. Seattle: Mountaineers Press.

Jones, Roy. 1972. *Wappato Indians: Their History and Prehistory*. Privately printed.

Juntunen, Judy Rycraft, May Dasch, and Ann Bennett Rogers. 2005. *The World of the Kalapuya*. Philomath, Ore.: Benton County Historical Society and Museum.

Kallas, John. Wapato: Indian potato. *Wilderness Way* 9(1). Available via www.wwmag.net/wapato.htm. Accessed 9 April 2008.

Leopold, Aldo. 1949. *A Sand County Almanac, and Sketches Here and There*. New York: Oxford University Press.

Leopold, Luna B., ed. 1953. *Round River: From the Journals of Aldo Leopold*. New York: Oxford University Press.

Long, Kim. 2000. *Beavers: A Wildlife Handbook*. Boulder, Colo.: Johnson Books.

Lopez, Barry, ed. 2006. *Home Ground: Language for an American Landscape*. San Antonio, Tex.: Trinity University Press.

Mackey, Harold. 2004. *The Kalapuyans*. Salem, Ore.: Mission Hill Museum Association.

Marshall, David, Matthew Hunter, and Alan Contreras. 2003. *Birds of Oregon: A General Reference*. Corvallis: Oregon State University Press.

Mason, Bill. 1984. *The Path of the Paddle*. Buffalo, N.Y.: Firefly Books.

Mason, Bill. 1988. *Song of the Paddle*. Buffalo, N.Y.: Firefly Books.

Moore, Lucia, Nina McCornack, and Gladys McCready. 1949. *The Story of Eugene*. New York: Stratford House.

Morrison, Dorothy Nafus. 1999. *Outpost: John McLoughlin and the Far Northwest*. Portland: Oregon Historical Society Press.

Moulton, Gary E., ed. 1991. *The Journals of the Lewis and Clark Expedition*. Vol. 7. Lincoln: University of Nebraska Press.

Natural Resources Conservation Service. Bull trout *Salvelinus confluentus* fact sheet. Available via www.mt.nrcs.usda.gov/news/factsheets/bulltrout.html. Accessed 1 May 2008.

Nedeau, Ethan, Allan K. Smith, and Jen Stone. 2005. *Freshwater Mussels of the Pacific Northwest*. U.S. Fish and Wildlife Service, Columbia River Fisheries Program Office, Vancouver, Wash.

Olson, Sigurd. 1956. *The Singing Wilderness*. New York: Alfred A. Knopf.

Oregon Department of Fish and Wildlife. Native Fish Investigations Project: Oregon chub. Available via http://oregonstate.edu/dept/ODFW/NativeFish/OregonChub.htm. Accessed 30 March 2008.

Pojar, Jim, and Andy MacKinnon, eds. 1994. *Plants of the Pacific Northwest Coast*. Vancouver, B.C.: Lone Pine.

Robbins, William G. 2004. *Landscapes of Conflict: The Oregon Story 1940–2000*. Seattle: University of Washington Press.

Rubin, Rick. 1999. *Naked Against the Rain: The People of the Lower Columbia River*. Portland, Ore.: Far Shore Press.

Snyder, Eugene. 1970. *Early Portland: Stump Town Triumphant*. Portland, Ore.: Binford & Mort.

Starbird, Ethel. 1972. A River Restored: Oregon's Willamette. *National Geographic* 141(6): 816–835.

Turner, Mark, and Phyllis Gustafson. 2006. *Wildflowers of the Pacific Northwest*. Portland, Ore.: Timber Press.

U.S. Fish and Wildlife Service. Species fact sheet: Oregon chub. Available via www.fws.gov/oregonfwo/Species/Data/OregonChub. Accessed 28 March 2008.

Walth, Brent. 1994. *Fire at Eden's Gate*. Portland: Oregon Historical Society Press.

Wentz, Dennis, Bernadine Bonn, and Kurt Carpenter. 1998. *Water Quality in the Willamette Basin, 1991–95*. U.S. Geological Survey Circular 1161. Portland, Ore.: U.S. Department of the Interior.

Williams, Travis. 2002. *Citizen's Guide to the Portland Harbor Cleanup*. Portland, Ore.: Willamette Riverkeeper.

Williams, Travis. 2007. *Willamette River Water Trail Guide: Buena Vista Ferry to the Columbia River*. Portland, Ore.: Willamette Riverkeeper.

Wood, Mary Christina. 2007. Nature's Trust: A Legal, Political, and Moral Frame for Global Warming. Paper presented at the 25th Annual Public Interest Environmental Law Conference, March 1–4, University of Oregon School of Law, Eugene.

Worth, Veryl M., and Harry S. Worth. 1989. *Early Days on the Upper Willamette*. Oakridge, Ore.: Fact Book Co.

Xerxes Society. 2008. Fender's blue butterfly page on the Butterfly Conservation Initiative website. http://www.butterflyrecovery.org/species_profiles/fenders_blue/. Accessed 12 April 2008.

Index

About the Author

TRAVIS WILLIAMS has worked in river conservation since the 1990s and since 2000 has led Willamette Riverkeeper, a nonprofit organization that focuses on clean water, habitat restoration, and low-impact river recreation, principally in canoes and kayaks. Earlier he worked for American Rivers and Conservation International in Washington DC. He is an avid canoeist who has traveled many western rivers and photographed their natural beauty. He holds a B.A. in International Studies from Portland State University and an M.S. in Environmental Science from The Johns Hopkins University. A fifth-generation Oregonian who grew up in Milwaukie, Oregon, he was on the Willamette River with friends at a young age. He can often be found paddling the Willamette, as well as other western rivers, with his two daughters and his partner, Sandra. Among his biggest priorities for river restoration is to reconnect side channels and floodplains to the Willamette River.